USING HPC FOR COMPUTATIONAL FLUID DYNAMICS

USING HPC FOR COMPUTATIONAL FLUID DYNAMICS

A Guide to High Performance Computing for CFD Engineers

SHAMOON JAMSHED

AMSTERDAM • BOSTON • HEIDELBERG • LONDON
NEW YORK • OXFORD • PARIS • SAN DIEGO
SAN FRANCISCO • SINGAPORE • SYDNEY • TOKYO

Academic Press is an imprint of Elsevier

Academic Press is an imprint of Elsevier
225 Wyman Street, Waltham, MA 02451, USA
The Boulevard, Langford Lane, Kidlington, Oxford OX5 1GB, UK

Notices
Knowledge and best practice in this field are constantly changing. As new research and experience
broaden our understanding, changes in research methods, professional practices, or medical treatment
may become necessary.

Practitioners and researchers must always rely on their own experience and knowledge in evaluating
and using any information, methods, compounds, or experiments described herein. In using such
information or methods they should be mindful of their own safety and the safety of others,
including parties for whom they have a professional responsibility.

To the fullest extent of the law, neither the Publisher nor the authors, contributors, or editors, assume
any liability for any injury and/or damage to persons or property as a matter of products liability,
negligence or otherwise, or from any use or operation of any methods, products, instructions, or ideas
contained in the material herein.

ISBN: 978-0-12-801567-4

British Library Cataloguing-in-Publication Data
A catalogue record for this book is available from the British Library

Library of Congress Cataloging-in-Publication Data
A catalog record for this book is available from the Library of Congress

For information on all Academic Press publications
visit our website at http://store.elsevier.com/

Typeset by TNQ Books and Journals
www.tnq.co.in

Printed and bound in the United States of America

  Working together
to grow libraries in
developing countries

www.elsevier.com • www.bookaid.org

Publisher: Joe Hayton
Acquisition Editor: Brian Guerin
Editorial Project Manager: Natasha Welford
Production Project Manager: Lisa Jones
Designer: Matthew Limbert

DEDICATION

Dedicated to my family and beloved wife Afia

CONTENTS

PREFACE

This work is an effort to produce a text that is helpful for those who are becoming engineers, and for postgraduate students and information technology (IT) professionals. Without doubt, Computational Fluid Dynamics (CFD) is an emerging technology advancing with the arrival of modern supercomputers. Therefore, knowledge of CFD alone is not sufficient to compete with ongoing challenges in this field. Many engineers are handicapped when dealing with IT matters and they wait a long time for an IT expert to solve problems. Therefore, to a certain extent, knowledge of High-Performance Computing (HPC) is mandatory for engineers. From a different perspective, when IT professionals are dealing with CFD problems, they need to know the basics of most ongoing commercial codes, especially their installation on Linux-based clusters, which enables efficient job scheduling and troubleshooting bottlenecks in running CFD software without the help of CFD engineers. It is important that the reader not feel the need to browse the Internet for simple matters which this book attempts to explain more easily. This book thus focuses on two aspects: HPC and CFD.

In the initial chapters, we concentrate on CFD aspects, its rationale, and advanced numerical schemes. Later chapters focus on HPC and the way HPC works in CFD, and then switch to different software variations as they are used in HPC, with a particular focus on commercial code ANSYS Fluent benchmarks.

Chapter 1 introduces CFD. Many organizations implement CFD in the computer-aided engineering phase. However, most of the time, higher management is not interested, perhaps because of lengthy simulations or the uncertainty of results. These issues are discussed and various misconceptions about CFD are explored and clarified. The basics of CFD with governing equations are also discussed.

Chapter 2 introduces HPC. The pioneers of HPC and their contributions are discussed and the world Top 5 computers are mentioned and discussed in detail.

Chapter 3 focuses on CFD algorithms designed for parallel machines. This is basically the treatment of CFD codes for parallel processing, because CFD codes designed for traditional serial machines may not run efficiently on parallel computers. This chapter focuses on how CFD can be used in parallel and on parallel architectures, with some discussion of mesh partitioning techniques.

Chapter 4 discusses turbulence and its high Reynolds representations, such as Direct Numerical Simulations (DNS), Large Eddy Simulations (LES), and Detached Eddy Simulations. The methods are not discussed from a mathematical point of view, to maintain the reader's interest. Some examples of DNS and LES are given and discussed: particularly, free shear flows and wall-bounded flows. Computational requirements in floating point operations per second with respect to the Reynolds number are mentioned, which is certainly useful if the user wants an idea of cost estimation for his or her machine, keeping in mind the size of the problem. Detached eddy simulation rationales are touched on and not discussed in detail because the method is not sensitive to very high computational power compared with LES and DNS. The chapter concludes with a discussion on the importance of each method.

Chapter 5 contains mainly a classification of clusters. Many types of clusters are used globally. It all depends on the needs of the user and the budget the solution requires. This chapter gives an idea about different types of distributed clusters and their advantages. The qualities of clusters and the different components of clusters necessary to construct an HPC system are also discussed in detail. The details of chassis-based servers and rack-mounted and desktop-based clusters such as CRAY CX1 are also discussed.

Chapter 6 is fairly large chapter that deals mainly with the onboard usage of HPC in commercial CFD codes. This is a detailed chapter that explains the memory requirement for ANSYS Fluent based on the problem size, and then interconnectivity, which has a vital role in HPC, is discussed with a focus on Infiniband. Storage requirements are also highlighted. After that, Fluent benchmarks are discussed with benchmarking of medium- to high-level problems. Other codes such as ANSYS CFX and OpenFOAM are also described from a benchmark point of view. A flowchart is provided so that the user can easily purchase or form an HPC machine. A comparison of different types of machines is included to give the user an idea as to how a machine would be suitable for his or her needs.

The performance of world-famous supercomputer t-Platforms is discussed. It is a leading HPC company in Russia. With headquarters in Germany, the company is credited for establishing one of the biggest Russian clusters, named Lomonosov, at Moscow State University. I mention and discuss in detail the Lomonosov cluster and its setup, high-power availability, and performance. In addition to Lomonosov, small clusters of t-Platforms are mentioned and described briefly.

Chapter 7 discusses HPC from a networking point of view. Transmission Control Protocol/Internet Protocol, Internet Protocol addressing, ssh, and remote access via PuTTy and WinSCP for running Fluent jobs are discussed in detail. Configuring Fluent on an Windows HPC server is described.

The last chapter discusses Graphics Processing Units (GPUs), an emerging technology in the HPC and graphics fields. Apart from graphics support, GPUs are able to perform routine calculations like an ordinary CPU, but several times faster and in a more efficient and optimized way. The CUDA architecture, which acts like the life blood of GPUs, is discussed in detail in this chapter. Two generations of NVIDIA GPUs, NVIDIA Tesla and GT200, are covered. Finally, with a focus on CFD applications, the use of GPU in CFD is discussed, mentioning the research of different scientists and engineers. This effort has been done to facilitate the work of engineers, scientists, and IT professionals, to serve as a meaningful and interesting doctrine in the field of CFD and HPC.

I hope that this text will be beneficial for the CFD engineers, scientists, and HPC professionals who are keen to learn about HPC and CFD and link the two. High-performance computing is the present and the future; one must be familiar with all of its basic needs and prerequisites. It surely has room for improvement, and the reader's response is highly welcome, but it is also hoped that this text will quench the thirst of those who want to excel in both research areas.

Shamoon Jamshed
Karachi, December 2014

CHAPTER 1

Introduction to CFD

1. COLORFUL DYNAMICS OR COMPUTATIONAL FLUID DYNAMICS?

Computational fluid dynamics (CFD) is one of the most quickly emerging fields in applied sciences. When computers were not mature enough to solve large numerical problems, two methods were used to solve fluid dynamics problems: analytical and experimental. Analytical methods were limited to simplified cases such as solving one-dimensional (1D) or 2D geometry, 1D flow, and steady flow. However, experimental methods demanded a lot of resources such as electricity, expensive equipment, data monitoring, and data post-processing. Sometimes for engineering analysis work, it is not within the budget of a small organization to establish such a facility. However, with the advent of modern computers and supercomputers, life has become much easier. With the passage of time numerical methods got matured and are now used to solve complex fluid dynamics problems in a short time. Thus, today, with a small investment, some good configuration personal computers can be bought and used to run CFD code that can handle complex flow geometries easily. The results can be achieved more quickly if some of the computers are joined or clustered together.

From an overall perspective, CFD is more economical than experiments. The twentieth century has seen the computer age move with cutting-edge changes, and problems or experiments that had never been thought possible to be performed experimentally or were difficult to perform because of limited resources are now possible with the modern technology. It can be said that CFD is more economical than experiments. With the advent of modern computer technology, it has gained in popularity as well because advanced methods for solving fluid dynamics equations can be analyzed quickly and efficiently.

In terms of accuracy, CFD lies in between the domain of theory and experiments. Because experiments mostly replicate real phenomena, they are much reliable. Analytical method is second because of certain assumptions involved while solving a particular problem. CFD is last because of it involves truncation errors, rounding off errors, and machine errors in

1

numerical methods. To avoid making it "colorful dynamics," it is the responsibility of the CFD analyst to fully understand the logic of the problem and correctly interpret results.

There are many benefits to performing CFD for a particular problem. A typical design cycle now contains two and four wind-tunnel tests of wing models instead of the 10–15 that were once routine. Because our main focus is High-Performance Computing (HPC), we can say that if CFD is the rider, HPC is the ride. Through HPC complex simulations (such as very high-speed flow) are possible that otherwise would have required extreme conditions for a wind tunnel. For hypersonic flow in the case of a re-entry vehicle, for example, the Mach number is 20 and CFD is the only viable tool with which to see flow behavior. For these vehicles, which cross the thin and upper atmosphere levels, nonequilibrium flow chemistry must be used.

Consider the example of a jet engine whose entire body is filled with complex geometries, faces, and curvature. CFD helps engineers design the after-burner mixers, for example, which provide additional thrust for greater maneuverability. Also, it is helpful in designing nacelles, bulbous, cylindrical engine cowlings, and so forth.

2. CLEARING MISCONCEPTIONS ABOUT CFD

An obvious question is why so many CFD users seem unhappy. Sometimes the problem lies in beliefs regarding CFD. Many organizations do not place value on CFD and rely on experiments. According to their view, the use of experiments is customary even though experiments are also prone to errors.

In addition, CFD has captured the research market quicker than experiments owing to the worldwide economic crisis, and it is the obvious choice over experiments for a company when a sufficient budget is scarce. It is also unfortunate that many people do not trust CFD, including the heads of companies and colleagues who sometimes do not understand the complexity of fluid dynamics problems. The analyst must first dig for errors, if any, and then examine how he or she should portray it to higher management. If management is spending money buying expensive hardware and software and hiring people, the importance of CFD is clear. If management still does not recognize the importance of CFD facts, it becomes the job of analysts to educate and mentor the bosses. If it is desired that the statements/arguments related to CFD remain unquestioned, they must be provided either with some scientific or mathematical proof or with some acknowledgment by those who have understanding and firm believe in the truth of the results.

One should compromise for less reliable CFD results when it is known that not enough computational resources are available. This brings us to a question regarding the control of uncertainties. Certain numerical schemes result in dissipation error, such as first order. Other schemes such as second-order result in dispersion error. Then there is machine error, grid accuracy error, human error, and truncation error, to name a few. Thus, unexpected predictions could cause the question, "Did I do something wrong?." In this case, it is essential to familiarize the user completely with CFD tool(s) and avoid allowing him or her to use the tool as a black box.

Many engineers do not pursue product development, design, and analysis as deeply as do CFD engineers. They do not understand turbulence modeling, convergence, mesh, and such. To sell something in the market using CFD, one should be smart and clever enough to say something the customer can understand.

It is also annoying when software does not correspond the way it should. This occurs when results do not converge or when there is some complex mesh to deal with. At first, one should:

1. Carefully make assumptions if required.
2. Try to make the model simpler (such as using a symmetric or periodic boundary condition).
3. Use reasonable boundary conditions. With an excellent mesh, results do not converge mostly owing to incorrect boundary conditions.
4. Monitor convergence.
5. If not satisfied, go to mesh.
6. If experimental data are unavailable, perform a grid convergence study.

In this way, the efforts will not change skeptics' perceptions overnight but if a history of excellent CFD solutions is delivered, they will start to believe it.

Although CFD has been criticized, there are many great things about it. A CFD engineer enjoys writing code and obtaining results, which increases his confidence level. From a marketing point of view, people are mostly attracted to the colorful pictures of CFD, which is how one can make a presentation truly overwhelming. If one can produce good results but cannot present the work convincingly, then all of the effort is useless.

From this discussion, it can be concluded that there are two important points to remember. One is that the problem does not lie in CFD but could be in the limitation of resources, lack of experimental data, or wrong interpretation of results. Second, skepticism regarding CFD exists but one should be smart enough to present the results in an attractive and evocative manner. Remember the saying that a drop falling on a rock over a long time can create a hole in it. That philosophy will definitely work here, as well. CFD can be colorful dynamics or computational fluid dynamics with colorful, meaningful results. It is your choice: What do you want to see and what do you see?

3. CFD INSIGHT

CFD mainly deals with the numerical analysis of fluid dynamics problems, which embodies differential calculus. The equations involved in fluid dynamics are Navier–Stokes equations. Until now, solutions to Navier–Stokes equations have not been explicitly found except for some cases such as Poiseuille flow, Couette flow, and Stokes flow with certain assumptions. Therefore, several engineers and scientists have spent their lives devising methods to solve these differential equations so as to give a meaningful solution for a particular set of geometry and initial conditions. Thus, CFD is the process of converting the partial differential equations of fluid dynamics into simple algebraic equations and then solving them numerically to obtain some meaningful result.

3.1 Comparison with Computational Structure Mechanics

Because it is a numerical tool, CFD relies heavily on experimental or analytical data for validation. In the author's experience, people who are in the field of computational structure mechanics (CSM) using Finite Element Analysis (FEA) codes for structural deformation in solids do not bother much about creating the grid. This is because the field of FEA is more mature than CFD. For example, there are no complex issues to solve such as the boundary layer, so meshing efforts are reduced. No monster exists such as y+, so life is easier.

In addition, CFD and CSM have two features in common: they both require meshing and they both require HPC when the mesh size is increased. In FEA, as the mesh is increased or the number of nodes increases, the size of the matrices to be solved increases. Similarly, when CFD problems are solved, the number of iterations or calculations increases with the number of grid points, which ultimately need more computational power.

3.2 CFD Process

The entire CFD process consists of three stages: pre-processing, solving, and post-processing. These are diagrammed in Figure 1.1.

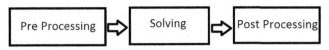

Figure 1.1 The computational fluid dynamics process.

All three processes are interdependent. As much as 90% of effort is used in the meshing (preprocessing) stage. This requires the user to be dexterous and there must be the idea of creating an understandable topology. The next stage is to solve the governing equations of flow, which is the computer's work. Remember that an error embedded in the mesh will propagate in the solving stage as well, and if you are lucky enough, you may get a converged solution. However, mostly, owing to only one culprit cell, the solution diverges. The next phase after solving equations is post-processing. There, the results of whatever was input and solved are obtained; colorful pictures showing contours are interpreted for product design, development, or optimization. For validation, the results are compared with experimental data. If any experimental data are absent, the grid convergence study better judges the authenticity of the results. In that case, the mesh is refined two or three times, each time solving and getting results, until a never-changing result (asymptotically converged solution) is obtained.

Post-processing has its own delights, and you can impress people by showing flow simulations such as path lines, flow contours, vector plots, flow ribbons, cylinders, and so forth. In unsteady flows, such as for direct numerical simulation (DNS) and large eddy simulation (LES), the iso-surface of Q-criterion or λ-criterion is also shown sometimes. Post-processing software such as Tecplot has the ability to see multiple things simultaneously in a single picture. As examples, the stream line and flow contours are shown simultaneously in Figure 1.2 for Ariane5 base flow [2] and Figure 1.3 [1] shows the flow over a delta wing. There, the iso-surface of constant pressure is shown over the wing, which is colored by the Mach number. An iso-surface is a surface formed by a collection of points with the same value of a property (such as temperature pressure).

We will focus on turbulent flows in this text because these problems are mostly solved with HPC machines. Turbulence is caused by instabilities in flow and is nondeterministic. There is a range of scales in turbulent flows that can be as large as half the size of the body that causes the turbulence or smaller than one-tenth of a millimeter. In current CFD techniques, in which we use mesh to solve the flow, the mesh size must be such that it contains cells not larger than the size of the smallest scale. Thus, the total mesh size

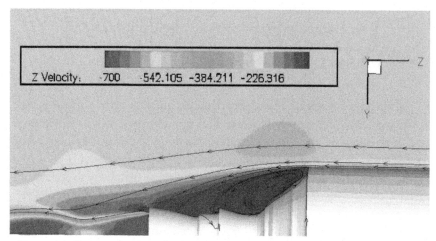

Figure 1.2 Flow structure at the base of Ariane5 ESA Satellite Launch Vehicle [2].

becomes too large for these problems: It could be over 30 million cells. A case for benchmarking such a problem is discussed in Chapter 6, where the mesh size is 111 million cells. The question is where to solve them. That is the purpose of this book, and why we need HPC. HPC solves these problems for us. Figure 1.4 shows an image of the vortices formed behind a truck body. The vortices obviously form as a result of the turbulence in the wake region. A corridor of low velocity often forms behind bodies, called

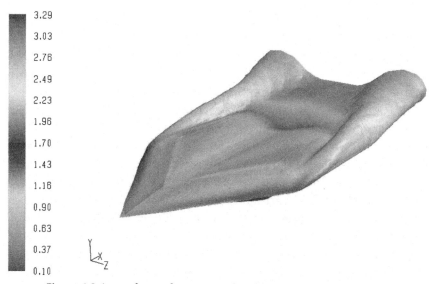

Figure 1.3 Iso-surfaces of pressure colored by the Mach number [1].

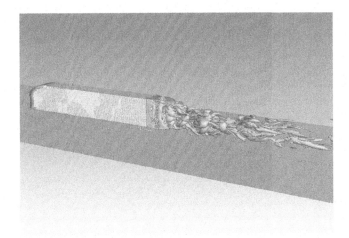

Figure 1.4 Eddies form and computed in the wake owing to the technique of Detached Eddy Simulation (DES), which cannot be seen with conventional Reynolds Averaged Navier Stokes (RANS) methods.

the wake. The scales I was talking about are the length of these vortices, often called eddies. These eddies frequently form and are miscible near the wall of the truck, whereas far from the truck body they mix with outside air and dissipate in the form of heat. If exhaust from the truck is also considered, a more realistic flow would be formed but computationally it would be more complex to solve. An attractive picture of vortices is shown in Figure 1.4.

3.3 Governing Equations

While we are talking about CFD, the discussion is incomplete without mentioning the governing equations. These equations are the life blood of CFD. The famous equations of fluid dynamics are also known as the Navier–Stokes equation. These equations were discovered independently more than 150 years ago by the French engineer Claude Navier and the Irish mathematician George Stokes. Application of supercomputers to solve these equations introduced the field of CFD. The basis of these equations lies in the assumption that a fluid particle deforms under shear stress. Then, using the second law of motion and energy conservation, the dynamics of the particle is described by its mass, momentum, and energy. In principle, all three parameters must be conserved:

1. Conservation of mass
2. Conservation of momentum
3. Conservation of energy

These set of equations constitute the Navier–Stokes equations. We will not explore the derivation of each of these equations. Instead, the equations are mentioned and complex terms are elaborated.

3.3.1 Conservation of Mass

The conservation of mass equation says that the mass is conserved. It is also known as the continuity equation. The continuity equation is written as:

$$\frac{\partial \rho}{\partial t} + \nabla \cdot \left(\rho \vec{U}\right) = 0 \tag{1.1}$$

where

$$\vec{U} = [u, v, w]$$

3.3.2 Conservation of Momentum

The conservation of momentum is based on Newton's second law that $F = ma$, where m is the mass of the fluid particle and a is its acceleration.

When applied to a fluid particle under the action of pressure, viscous and body forces constitute a set of momentum equations. In the x-direction,

$$\frac{\partial \rho u}{\partial t} + \frac{\partial \rho uu}{\partial x} + \frac{\partial \rho uv}{\partial y} + \frac{\partial \rho uw}{\partial z} = -\frac{\partial p}{\partial x} + \mu \frac{\partial}{\partial x}\left(\frac{\partial u}{\partial x} + \frac{\partial u}{\partial y} + \frac{\partial u}{\partial z}\right) + \rho f_x \tag{1.2a}$$

In the y-direction,

$$\frac{\partial \rho v}{\partial t} + \frac{\partial \rho vu}{\partial x} + \frac{\partial \rho vv}{\partial y} + \frac{\partial \rho vw}{\partial z} = -\frac{\partial p}{\partial y} + \mu \frac{\partial}{\partial y}\left(\frac{\partial v}{\partial x} + \frac{\partial v}{\partial y} + \frac{\partial v}{\partial z}\right) + \rho f_y \tag{1.2b}$$

and in the z-direction,

$$\frac{\partial \rho w}{\partial t} + \frac{\partial \rho wu}{\partial x} + \frac{\partial \rho wv}{\partial y} + \frac{\partial \rho ww}{\partial z} = -\frac{\partial p}{\partial z} + \mu \frac{\partial}{\partial z}\left(\frac{\partial w}{\partial x} + \frac{\partial w}{\partial y} + \frac{\partial w}{\partial z}\right) + \rho f_z \tag{1.2c}$$

In 1845, Stokes obtained a relation for the shear stress of Newtonian fluids. With some modification in the viscosity terms, these equations are given as follows:

$$\tau_{xx} = \lambda\left(\nabla \cdot \vec{U}\right) + 2\mu \frac{\partial u}{\partial x} \tag{1.2d}$$

$$\tau_{yy} = \lambda\left(\nabla \cdot \overrightarrow{U}\right) + 2\mu\frac{\partial v}{\partial y} \tag{1.2e}$$

$$\tau_{zz} = \lambda\left(\nabla \cdot \overrightarrow{U}\right) + 2\mu\frac{\partial w}{\partial z} \tag{1.2f}$$

Asymmetric stress tensors are given as:

$$\tau_{xy} = \tau_{yx} = \mu\left(\frac{\partial v}{\partial x} + \frac{\partial u}{\partial y}\right) \tag{1.2g}$$

$$\tau_{yz} = \tau_{zy} = \mu\left(\frac{\partial v}{\partial z} + \frac{\partial w}{\partial y}\right) \tag{1.2h}$$

$$\tau_{zx} = \tau_{zx} = \mu\left(\frac{\partial w}{\partial x} + \frac{\partial u}{\partial z}\right) \tag{1.2i}$$

where μ is the dynamics viscosity and λ is the second viscosity coefficient, given by Stokes as:

$$\lambda = -\frac{2}{3}\mu \tag{1.2j}$$

3.3.3 Energy Equation

The energy equation is based on the principle that energy is conserved. It is also called the First Law of Thermodynamics. Its general form is given in Eqn (1.3):

$$\frac{\partial(\rho E)}{\partial t} + \nabla \cdot \left(\rho E\overrightarrow{U}\right) = \rho\dot{q} + \frac{\partial}{\partial x}\left(k\frac{\partial T}{\partial x}\right) + \frac{\partial}{\partial y}\left(k\frac{\partial T}{\partial y}\right) + \frac{\partial}{\partial z}\left(k\frac{\partial T}{\partial z}\right)$$

$$- p\nabla \cdot \overrightarrow{U} + \lambda\left(\nabla \cdot \overrightarrow{U}\right)^2 + \mu\left[2\left(\frac{\partial u}{\partial x}\right)^2 + 2\left(\frac{\partial v}{\partial y}\right)^2\right.$$

$$+ 2\left(\frac{\partial w}{\partial z}\right)^2 + \left(\frac{\partial u}{\partial y} + \frac{\partial v}{\partial x}\right)^2 + \left(\frac{\partial v}{\partial z} + \frac{\partial w}{\partial y}\right)^2$$

$$\left. + \left(\frac{\partial w}{\partial x} + \frac{\partial u}{\partial z}\right)^2\right]$$

$$\tag{1.3}$$

These three governing equations, i.e., continuity, momentum, and energy, constitute the Navier–Stokes equations. Some authors refer to only momentum equations as Navier–Stokes equations. These equations are important in CFD and the reader should memorize them to understand the methods of CFD. In CFD these equations are discretized along with points in space and then solved algebraically.

To solve these, various approaches are used, such as the finite difference method (FDM), finite volume method (FVM), and finite element method. These equations can be modified for inviscid, incompressible, or compressible and steady or unsteady fluid flow. For an inviscid flow field, the viscous terms would be neglected and the leftover equations would then be referred to as Euler equations.

In theory, the Navier–Stokes equations describe the velocity and pressure of fluid accelerating by any point near the surface of a body. If we consider an aircraft body as an example; these data can be used by engineers to compute, for various flight conditions, all aerodynamic parameters of interest, such as the lift, drag, and moment (twisting forces) exerted on the airplane. Drag is particularly important with respect to the fuel efficiency of an aircraft because it is one of the largest operating expenses for most airlines. It is not surprising that many aircraft companies spend a large amount of money for drag reduction research even if it results in one-tenth of a percent. Computation-wise, drag is the most difficult to compute compared with moment and lift.

To make these equations understandable to a computer, it is essential to represent the aircraft's surface and the space around it in a form that is usable by the computer. To do this, codes are developed in which the aircraft and its surroundings are represented as a series of regularly spaced points called a computational grid. These are then supplied to the solver code that applies Navier–Stokes equations to the grid data. The computer then computes the values of air velocity, pressure, temperature, and so forth, at all points. In effect, the computational grid breaks up the computational problem in space; the calculations are carried out at regular intervals to simulate the passage of time, so the simulation is temporally discretized as well. The closer the gird points are, the more often they will be computed and the more accurate and realistic the simulation is.

The problem is still not straightforward. The Navier–Stokes equations are in fact nonlinear, so many variables in these equations vary with respect to each other by powers of two or greater. Interaction of these nonlinear variables creates newer terms, which makes the solution difficult to solve. In addition, the global dependence of variables augments the complexity, such as the pressure which at a point depends on the flow at many other points. Because the different parts of a single problem are so intermingled, the solution must be obtained at many points simultaneously.

While we are dealing with CFD, keep in mind that our main focus in the CFD area is turbulence. This is because turbulence is currently the door for which HPC is the key. Only computational solutions give a detailed prediction of turbulence, which is not possible through other experimental or analytical means.

4. METHODS OF DISCRETIZATION

There are several methods of discretization that are programmed in commercial codes. ANSYS FLUENT and ANSYS CFX both use FVM. This is because FVM has certain advantages and the scheme is robust. The most popular methods are FDM and FVM, and we will discuss them next.

4.1 Finite Difference Method

Of all methods, FDM is the simplest. It can be said that CFD started from FDM. Initially, mathematicians derived simple formulas to calculate derivatives and then the methods improved and CFD advanced to more advanced methods. Currently, computations such as DNS and LES are only theoretical. The rationale of FDM can be understood from the concept of a derivative. The derivative of a function gives the slope of the function. For a function of x-component of velocity u, the slope of u with respect to x can be determined numerically as:

$$\frac{\partial u}{\partial x} = \frac{u_{i+1} - u_i}{\Delta x} \qquad (1.4)$$

where the subscripts i and $i + 1$ are the points for calculating the u values. Here, Δx denotes the grid spacing. The method for calculating the first derivative is also called the forward difference method, as we will soon observe.

4.2 Taylor Series Expansion and Forward Difference
4.2.1 Forward Difference Scheme
All of the derivatives are derived numerically using Taylor series expansion. The number of points used to evaluate the derivative is called the stencil.

The higher the number of points is, the more accurate will be the numerical result. Consequently, the spacing reduction between points also improves accuracy, but with some limitations. For a forward difference approximation the stencil would be i and $i + 1$. Hence, for function u the value at $i + 1$, i.e., u_{i+1}, would be:

$$u_{i+1} = u_i + \frac{\partial u}{\partial x}\Delta x + \frac{\partial^2 u}{\partial x^2}\left(\frac{\Delta x}{2!}\right)^2 + \frac{\partial^3 u}{\partial x^3}\left(\frac{\Delta x}{3!}\right)^3 + \cdots \qquad (1.5)$$

This formulation is then modified to evaluate the derivative. Divide each term by Δx and manipulate the terms:

$$\frac{\partial u}{\partial x} = \frac{u_{i+1} - u_i}{\Delta x} + O(\Delta x) \qquad (1.6)$$

This equation is called the first-order forward difference scheme. The last term, $O(\Delta x)$, indicates that the formulation is first order. This means that when the Taylor series formula was divided by Δx, the least derivative term containing the Δx term was of the order of 1 (power of 1).

4.2.2 Backward Difference Scheme

The backward difference scheme is as straightforward as the forward difference scheme. The stencil is reversed, i.e., i and $i - 1$. The Taylor series expansion is:

$$u_{i-1} = u_i - \frac{\partial u}{\partial x}\Delta x + \frac{\partial^2 u}{\partial x^2}\left(\frac{\Delta x}{2!}\right)^2 - \frac{\partial^3 u}{\partial x^3}\left(\frac{\Delta x}{3!}\right)^3 + \cdots \qquad (1.7)$$

Divide each term by Δx and manipulate the terms:

$$\frac{\partial u}{\partial x} = \frac{u_i - u_{i-1}}{\Delta x} + O(\Delta x) \qquad (1.8)$$

It should be noted that forward and backward schemes are first-order accurate.

4.2.3 Central Difference Scheme

The central differencing scheme is obtained by combining the forward series and backward series. This is done by subtracting the backward from the forward scheme. Subtracting Eqn (1.7) from Eqn (1.5), the following formulation is obtained:

$$\frac{\partial u}{\partial x} = \frac{u_{i+1} - u_{i-1}}{2\Delta x} + O(\Delta x) \qquad (1.9)$$

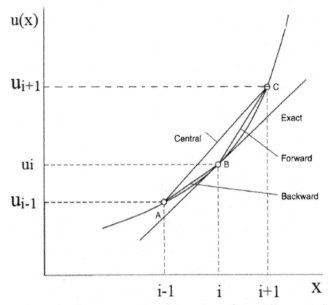

Figure 1.5 Graphical representation of three basic finite difference schemes.

The last term indicates the order of the central difference scheme. The central difference scheme is second-order accurate. Therefore, it is widely used in the calculations. The error that arises owing to the truncation (cutting) of the higher-order derivatives of the Taylor series terms is called the truncation error.

4.2.4 Second-Order Derivative

To obtain the second-order derivative, the method is slightly tedious. Because it involves more and more terms, the equations become more and more complex mathematically. By adding Eqns (1.5) and (1.7), we get:

$$\frac{\partial^2 u}{\partial x^2} = \frac{u_{i+1} - 2u_i + u_{i-1}}{\Delta x^2} + O(\Delta x^2) \tag{1.10}$$

The three methods mentioned above are shown in Figure 1.5.

4.3 Finite Volume Method

The FVM is widely used in CFD codes because of its various advantages over FDM. The FVM can be used for any sort of grid, i.e., structured or unstructured, clustered or non-clustered, and so forth. It can also be used in cases where there is discontinuity in the flow where FDM fails to calculate.

In the FVM the computational domain is divided into a number of control volumes. The values are calculated at cell centers. The values of fluxes at the cell interface are determined through interpolation using the values at the cell centers. For each control volume an algebraic equation is obtained, and thus a number of equations appear that are then solved using numerical methods. The FVM should not be confused with geometric volume definition. It has nothing to do with physical volume. Both schemes, i.e., FDM and FVM, can be used in 2D and 3D flow fields.

The term "volume" refers to the fact that, to solve fluid dynamics equations, the domain is discretized using control volumes (which could be 2D, as well) instead of taking discrete points as for the FDM. This is also a paramount reason to accommodate unstructured grids in FVM. One disadvantage of the FVM method is that higher-order schemes greater than second order are difficult to handle in three dimensions. This is because of dual approximations: that is, interpolation between the cell centers and the interfaces and the integration of all surfaces.

4.3.1 Simplest Approach: Gauss's Divergence Theorem

We begin by considering Gauss's divergence theorem for a control volume. The mesh for FVM can be structured or unstructured. Figures 1.6 and 1.7 show the structured and unstructured mesh for FVM. The normal $\bar{n}$ represents the vector normal to the surface. The function ϕ can be any perimeter such as velocity, temperature, or pressure. The first-order derivative in x-direction is:

$$\frac{\partial \phi}{\partial x} = \frac{1}{\Delta V} \int_V \frac{\partial \phi}{\partial x} dV = \frac{1}{\Delta V} \int_A \phi dA^x = \frac{1}{\Delta V} \sum_{i=1}^{N} \phi_i A_i^x \qquad (1.11)$$

where ϕ_i is the variable value at the elemental surfaces and N denotes the number of bounding surfaces on the elemental volume. Equation (1.11) applies for any type of finite volume cell that can be represented within the numerical grid. For the structured mesh shown in Figure 1.6, N has a value of 4 because there are four bounding surfaces of the element. In 3D, for a hexagonal element, N becomes 6. Similarly, the first-order derivative for ϕ in the y-direction is obtained, which can be written as:

$$\frac{\partial \phi}{\partial y} = \frac{1}{\Delta V} \int_V \frac{\partial \phi}{\partial y} dV = \frac{1}{\Delta V} \int_A \phi dA^y = \frac{1}{\Delta V} \sum_{i=1}^{N} \phi_i A_i^y \qquad (1.12)$$

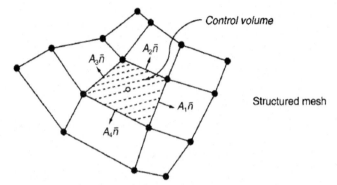

Figure 1.6 Structured mesh for finite volume method.

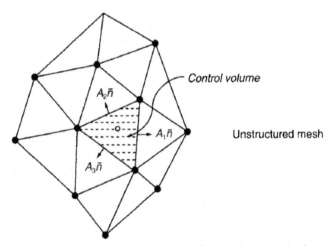

Figure 1.7 Unstructured mesh for finite volume method.

Problem

This problem describes the discretization of continuity equation using FVM. The results are shown and compared with the FDM solution of the same equation. The 2D continuity equation must be discretized:

$$\frac{\partial u}{\partial x} + \frac{\partial v}{\partial y} = 0 \tag{1.13}$$

on a structured grid.

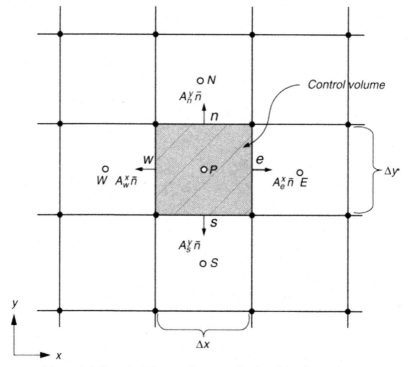

Figure 1.8 Stencil of finite volume method grid for the problem.

Solution: The stencil used for the problem is shown in Figure 1.8. Introducing control volume integration, that is, applying Eqns (1.11) and (1.12), yields the following expressions, which are applicable to both structured and unstructured grids:

$$\frac{\partial \phi}{\partial x} = \frac{1}{\Delta V} \sum_{i=1}^{N} \phi_i A_i^x = \frac{u_e A_e^x - u_w A_w^x + u_n A_n^x - u_s A_s^x}{\Delta V} \tag{1.14}$$

Similarly,

$$\frac{\partial \phi}{\partial y} = \frac{1}{\Delta V} \sum_{i=1}^{N} \phi_i A_i^y \tag{1.15}$$

For the structured uniform grid arrangement, the projection areas A_n^x and A_s^x in the x-direction and the projection areas A_e^y and A_w^y in the y-direction are zero. One important aspect demonstrated here by the FVM is that it allows direct discretization in the physical domain (or in a body-fitted conformal grid) without the need to transform the continuity equation from the physical domain to a computational domain as required in the FDM.

Because the grid has been considered to be uniform, face velocities u_e, u_w, v_n, and v_s are located midway between each of the control volume centroids, which allows us to determine the face velocities from the values located at the centroids of the control volumes. This implies:

$$u_e = \frac{u_E + u_P}{2}; \quad u_w = \frac{u_P + u_W}{2}; \quad v_n = \frac{v_N + v_P}{2}; \quad v_s = \frac{v_P + v_S}{2}$$

By substituting these expressions into the discretized form of the velocity first-order derivatives, the final form of the discretized continuity equation becomes:

$$\frac{u_E + u_P}{2}A_e^x - \frac{u_P + u_W}{2}A_w^x + \frac{v_N + v_P}{2}A_n^y - \frac{v_P + v_S}{2}A_s^y$$

Putting this into the above equation and manipulating, we get:

$$\left(\frac{u_E - u_W}{2\Delta x}\right) + \left(\frac{v_N - v_S}{2\Delta y}\right) = 0 \tag{1.16}$$

This is the continuity equation obtained with the FVM. It is interesting to check the results with the FDM. The above formulation has $2\Delta x$ and $2\Delta y$ in the denominator, which indicates the second–order accuracy of the scheme. If the central difference is used for the continuity equation which discretized just above, a similar formulation will be obtained. Using the stencil of Figure 1.8, the formulation can be made by applying the forward difference between E and P:

$$u_E = u_P + \frac{\partial u}{\partial x}\Delta x + \frac{\partial^2 u}{\partial x^2}\left(\frac{\Delta x}{2!}\right)^2 + \frac{\partial^3 u}{\partial x^3}\left(\frac{\Delta x}{3!}\right)^3 + \cdots \tag{1.17}$$

and the backward difference between P and W:

$$u_W = u_P - \frac{\partial u}{\partial x}\Delta x + \frac{\partial^2 u}{\partial x^2}\left(\frac{\Delta x}{2!}\right)^2 - \frac{\partial^3 u}{\partial x^3}\left(\frac{\Delta x}{3!}\right)^3 + \cdots \tag{1.18}$$

Because we know that the central difference is obtained by subtracting the backward from the forward formulation. Subtracting Eqn (1.18) from Eqn (1.17) implies:

$$\frac{u_E - u_W}{2\Delta x} \tag{1.19}$$

Similarly, for the y-component, performing this operation again with N and S cell centers and v as the velocity component, we get:

$$\frac{v_N - v_S}{2\Delta y} \qquad (1.20)$$

Summing both Eqns (1.19) and (1.20) and equating to zero to get the continuity equation:

$$\left(\frac{u_E - u_W}{2\Delta x}\right) + \left(\frac{v_N - v_S}{2\Delta y}\right) = 0 \qquad (1.21)$$

Comparing Eqn (1.21) with Eqn (1.16), there is no difference between them. Both schemes are second-order accurate.

5. TURBULENCE SHOULD NOT BE TAKEN FOR GRANTED

For a turbulent flow case, the variation in length scales, often represented by the ratio of the largest to the smallest eddy size, can be computed from the Reynolds number raised to the power 3/4. We will see in Chapter 4 that this ratio can be used to calculate the number of cells required for a reasonably accurate simulation. Because a practical problem is 3D, the number of grid points becomes proportional to the Reynolds number raised to the power 9/4. Thus, in simpler terms, doubling the Reynolds number increases the grid points five times.

Let us consider an aircraft that is 50 m long and wings with a chord length (the distance from the leading edge to the trailing edge) of about 5 m. It is cruising at 250 m/s at an altitude of 10 km. For this, 10^{16} grid points are needed to simulate turbulence near the surface with sufficient detail [3]. We will explore this in detail in Chapter 4, but for the time being note that currently, even with a supercomputer capable of performing 10^{12} floating point operations/second, it would take several thousand years to compute the flow for 1 s of flight.

In fact, engineers are rarely interested about detailed turbulent quantities such as small eddy dissipation; their main concern is usually with mean flow quantities, or in the case of aircraft, the lift and drag forces and heat transfer. In the case of an internal combustion engine (ICE), they could be interested in the rates at which the fuel and oxidizer mix. Including turbulence means including the fluctuating terms in the coupled Navier–Stokes equations. These fluctuating terms are mostly time-averaged, thereby smoothing the flow. This practice dramatically reduces the number of grid points. Some models used to model the terms arise as a result of the averaging procedure in Navier–Stokes. All of the models carry some assumptions and contain coefficients based on experimental findings. Therefore, it would be too good to be true to accept that a turbulent flow simulation could be as good as the model it contains.

6. TECHNOLOGICAL INNOVATIONS

Fluid dynamicists are able to find a way to directly simulate the greater portions of turbulent eddies with the advent of the fast computer era. In this way, there is a compromise between DNS and turbulence-averaged quantities simulation. The compromise has resulted in LES. With this technique, large eddies are resolved while small eddies are modeled. The technique emerged from meteorology, in which the large-scale turbulent motion of clouds was of particular interest. Currently, engineers are able to use the technique widely in other areas of fluid sciences such as for gases inside combustion chambers of ICE. DNSs are no longer theoretical. Simple cases such as pipe flow offer deep insight into turbulence through DNS. These simulations have also helped engineers fine-tune the coefficients used in models of turbulence.

A study was done by P. Moin [3,4] regarding the control of turbulence. The study was conducted on drag reduction for civilian aircraft. The concept of riblets was used, adopted from tooth-like structures on the skin surface of sharks. Numerical simulations using DNS showed that riblets tend to reduce the motion of eddies, preventing them from coming close to the surface of the body (within about 50 μm). This preserves the transportation of a high-speed fluid close to the surface. Another technology that employs this development is active control. Converse to the passive technology of riblets, this technology uses control surfaces to move against the turbulent fluctuations of an incoming fluid. The wing surface contains several micro-electromechanical systems that respond to pressure fluctuations to control the small eddies that cause turbulence. Fluid dynamicists are able to become involved in all of these efforts only by seeing nature in the form of a shark's skin movement. They are trying to build smarter aircrafts using this technique. This drag effect is not limited to aircraft skins; the technology is also used in golf balls. The drag exerted on the ball mostly results from pressure, which is more in front than behind the ball. Golf balls have dimples on their surface that increase turbulence and hence reduce drag, thereby increasing the ability to travel about two and half times farther than a plane-surface ball (Figure 1.9).

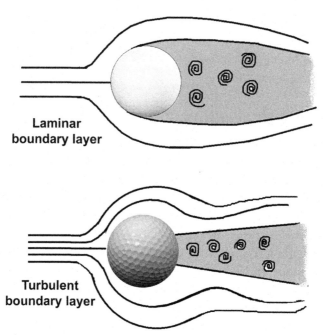

Figure 1.9 Flow behind a plane ball and a golf ball, a laminar boundary layer experiences more drag than a turbulent one.

Thus, CFD helps us in many complicated cases for which we cannot easily judge or analyze based on experimental or analytical data. The growing popularity of CFD is solely due to the rapid increase in computational power and the efficacy that is reflected by the field itself. However, because humanity's desires do not rest, as computational power crosses a quadrillion floating-point operations per second, scientists and fluid dynamicists will begin to float new complex problems that are currently thought to be impossible to solve.

REFERENCES

[1] Jamshed S, Hussain M. Viscous flow simulations on a delta rectangular wing using Spalart Allmaras as a turbulence model. In: Proceedings of the 8th international Bhurban conference on applied sciences & technology. (Islamabad, Pakistan): January 2011.
[2] Jamshed S, Thomber B. Numerical analysis of transonic base-flow of Ariane5. In: 7th International Bhurban conference on applied sciences & technology. (Islamabad, Pakistan): Centre of excellence on science and technology (CESAT); January 7, 2010.
[3] Moin P, Kim J. Tackling turbulence with supercomputers. Scientific American January 1997, pp 62–68.
[4] Moin P, Kim J, Choi H. Direct numerical simulation of turbulent flow over riblets. Journal of Fluid Mechanics October 1993;Vol. 255:503–39. http://dx.doi.org/10.1017/S0022112093002575.

CHAPTER 2

Introduction to High-Performance Computing

1. A SHORT HISTORY

"Necessity is the mother of invention" is a famous proverb. Science and technology have led us to a stage in which everything depends on computers. Modern computers are ubiquitous compared with earlier times. Everything is becoming smaller and smarter. Calculations that took half a year to solve now take a month; what took months now takes hours and what took hours now takes minutes. Innovations have also made size increasingly smaller. The big machines of 60–70 years ago, which were called "mainframes," can be seen on your desktop today. But sky is still the limit, and so after humans created big machines, they created big problems to solve using them. Because every problem has a solution, scientists and engineers have devised ways to run those problems; they have made the machines stronger and smarter and called the machine "supercomputers." Those giant monsters are state-of-the-art computers that use their magnificent power to solve big "crunchy" problems.

What is a supercomputer? It is a machine that has the highest performance rating at the time of its introduction into the world. But because the world changes dramatically quickly, this means that the supercomputer of today will not be a supercomputer tomorrow. Today, supercomputers are typically a kind of custom design produced by traditional companies such as Cray, IBM, and Hewlett–Packard (HP), which had purchased many of the 1980s companies to gain experience. The first machine generally referred to as a supercomputer was the IBM Naval Ordinance Research Calculator. It was used at Columbia University from 1954 to 1963 to calculate missile trajectories. It predated microprocessors, had a clock speed of 1 μs (1 MHz), and was able to perform about 15,000 operations per second. In the embryological years of supercomputing, the Control Data Corps (CDC's) early machines had very fast scalar processors, some 10 times the speed of the fastest machines offered by other companies at that time. Here, it would be unfair to neglect to mention the name of Seymour Cray (Figure 2.1),

Figure 2.1 Seymour Cray. *Courtesy of the history of computing project. http://www.tcop. net.*

who designed the first officially designated supercomputer for control data in the late 1960s. His first design, the CDC 6600, had a pipelined scalar architecture and used the RISC instruction set that his team developed. In that architecture, a single central processing unit (CPU) overlaps fetching, decoding, and executing instructions to process one instruction each clock cycle.

Cray pushed the number-crunching speed with the CDC 7600 before developing four-processor architecture with the CDC 8600. Multiple processors, however, raised operating system and software issues. When Cray left CDC in 1972 to start his own company, Cray Research, he ignored multiprocessor architecture in favor of vector processing. A vector is a single-column matrix, so a vector processor was able to implement an instruction or a set of instructions that could operate on one-dimensional arrays. Seymour then took over the supercomputer market with his new designs, holding the top spot in supercomputing for 5 years (1985–1990).

Throughout their early history, supercomputers remained the province of large governmental agencies and government-funded institutions. The production runs of supercomputers were small and their export was carefully controlled because they were used in critical nuclear weapons research. When the United States (US) National Science Foundation (NSF) decided to buy Japanese-made NEC supercomputer, it was believed to be a nightmare for US technology greatness. But later on, the world saw something different.

The early and mid-1980s saw machines with a modest number of vector processors working in parallel to become the standard. Typical numbers of processors were in the range of four to sixteen. In the later 1980s and 1990s, massive parallel processing became more popular than vector processors that contained thousands of ordinary CPUs.

Microprocessor speed found on desktops had overtaken the computing power of past supercomputers. Video games use the kind of processing power that had previously been available only in government laboratories. By 1997, ordinary microprocessors were capable of over 450 million theoretical operations per second. Today, parallel designs are based on server-class microprocessors such as the Xeon, AMD Opteron, and Itanium, and co-processors such as NVIDIA Tesla General Purpose Graphics Processing Unit, AMD GPUs, and IBM Cell. Technologists began building distributed and massively parallel supercomputers and were able to tackle operating system and software problems that had discouraged Seymour Cray from multiprocessing 40 years before. Peripheral speeds had increased so that input/output was no longer a bottleneck. High-speed communications made distributed and parallel designs possible. This suggests that the future of vector technology will not survive. However, NEC produced the Earth Simulator in 2002, which uses 5104 processors and vector technology. According to the top 500 list of supercomputers (http://www.top500.org), the simulator has achieved 35.86 trillion floating point operations per second (teraFLOPs).

Supercomputers are used for highly calculation-intensive tasks such as problems involving quantum physics, climate research, molecular modeling (computing the structures and properties of chemical compounds, biological macromolecules, polymers, and crystals), physical simulations (such as simulating airplanes in wind tunnels, simulating the detonation of nuclear weapons, and research on nuclear fusion).

Supercomputers are also used for weather forecasting, Computational Fluid Dynamics (CFD) (such as modeling air flow around airplanes or automobiles), as discussed in the previous chapter, and simulations of nuclear explosions—applications with vast numbers of variables and equations that have to be solved or integrated numerically through an almost incomprehensible number of steps, or probabilistically by Monte Carlo sampling.

2. HIGH-PERFORMANCE COMPUTING

High-performance computing (HPC) and supercomputing terms are not different. In fact, people use these terms interchangeably. Although the term "supercomputing" is still used, HPC is used much more frequently in the field of science and engineering. High-performance computing is a technology that focuses on the development of supercomputers (mainly), parallel processing algorithms, and related software such as Message Passing Interface and Linpack. High-performance computing is expensive but if we talk about CFD (which is our main focus), HPC is less costly than usual experimentation done in fluid dynamics. High-performance computing is needed in the areas of:

1. Research and development
2. Parallel programming algorithm development
3. Crunching numerical codes

Why do we need that? The answer to this question is simple. It is because we want to solve big problems. The question arises, "How big?" This really needs some justification, which we will explore. Computational fluid dynamics has been maturing in the past 50 years. Scientists want to solve flow, which requires the equations of fluid dynamics. As mentioned in the chapter on CFD, these equations (Navier–Stokes) are partial differential equations (PDEs), which we have been unable to solve completely until now. For this purpose, we use numerical methods, in which PDEs are broken down into a system of algebraic equations. Partial differential equations through numerical methods need a grid or mesh upon which to run. This is the basic root—the mesh, which solely depends on how powerful the machines are that you would need. If simple laminar flow is to be solved, a simple desktop personal computer (PC) will run the job (with a mesh size in the thousands). If a *Reynolds Averaged Navier–Stokes* (RANS) equation is needed, I can still use a desktop PC with its multiple cores. However, if we want to run unsteady (a time-dependent simulation), we might need a workstation and large storage. This situation is parallel to the case in which the mesh increases from 1 million cells (1 million is sufficient for RANS most of the time), when two workstations may be required. If more complexity is added like, for large eddy simulation, multiple workstations and a large amount of storage will be required. Thus, where to stop depends on our need and the budget the user would have to play with; between them, we can arrive at a reasonable solution. Computational fluid dynamics needs a lot of effort from both the user's end and the computational part. Scientists and engineers are trying to develop methods that take less effort to solve, are less prone to errors, and are computationally less expensive.

3. TOP FIVE SUPERCOMPUTERS OF THE WORLD

Top500 is a well-known Web site for the HPC community. It is updated annually in June and November by the National Energy Research and Scientific Computing Center, University of Tennessee in Knoxville, Tennessee, US, and the University of Mannheim in Germany. Top500 contains the list of the top 500 supercomputers. It is not possible to discussing all of them here. Interested readers can visit http://www.top500.org. In this chapter, the top five supercomputers will be discussed.

3.1 Tianhe 2 (Milky Way 2)

Tianhe 2 is the second generation of China's National University of Defense Technology's (NUDT's) first HPC, Tianhe 1 (*Tianhe* means "Milky Way"). Tianhe 2 became the world's number one system, with a peak performance of 33.86 petaFLOPs/s (PFLOP/s) of LINPACK benchmark [1]; in mid-2013. It has 16,000 compute nodes, each carrying two Intel Ivy Bridge Xeon processor chips and three Xeon Phi coprocessor chips. Thus, the total number of cores is 3,120,000. It is not known which model of the two processors was used: for example, the Intel Ivy Bridge Xeon processor is available with six, eight, or even 10 cores. Each of the 16,000 nodes possesses 88 GB memory and the total memory of the cluster is 1.34 PiB. Figure 2.2 shows the Tianhe 2 supercomputer.

Tianhe beat the Titan (the first-place holder before that), offering double peak performance. According to statistics from Top500, China houses 66 of the top 500 supercomputers, which makes it second compared

Figure 2.2 Tianhe 2 supercomputer. *Courtesy of Top500.*

with the US, which has 252 systems in the Top500 ranking. Without doubt, in the next 5 years most of the supercomputers in the Top500 market list will be owned by China.

3.1.1 Applications

The NUDT declared that Tianhe 2 will be used for simulations, analysis, and government security applications (cybersecurity, for example). After it was placed and assembled at its final location, the system had a theoretical peak performance of 54.9 PFLOP/s. The system draws about 17.6 MW of power for this performance. If the total amount of cooling is considered, this amount of power ends up at 24 MW. The system occupies 720 m^2 of space.

3.1.2 Additional Specs

The front-end node is equipped with 4096 Galaxy FT-1500 CPUs developed by NUDT. Each FT-1500 has 16 cores and 1.8 GHz clock frequency. The chip has a performance of 144 gigaFLOPs and runs on 65 W.

The interconnect is called TH Express-2 and was also designed by NUDT; and it has fat tree topology carrying 13 switches each with 576 ports.

The operating system (OS) is Kylin Linux. Because Linux has a GNU license it was modified by NUDT for their requirements. The Kylin OS was developed under the name Qilin, which is a mythical beast in China. This 14-year-old OS was initially developed for the Chinese military and other government organizations, but more recently, a version called NeoKylin that was announced in 2010 was developed. Linux resource management is done with Simple Linux Utility for Resource Management.

3.2 Titan, Cray XK7

Titan, a Cray XK7 system installed at the Department of Energy's (DOE) Oak Ridge National Laboratory, remains the number two system. It achieved 17.59 PFLOP/s on the Linpack benchmark whereas the theoretical performance was claimed to be 27.1 PFLOP/s. It is one of the most energy-efficient systems on the list, consuming a total of 8.21 MW and delivering 2.143 gigaFLOPs/s per watt (Figure 2.3). On the Green500 list, it ranks third in terms of the energy efficiency rating. It contains more than 18,000 nodes of KX7 with total AMD Opteron 6000 series CPUs with 32 GB of memory and K20X GPU with 6 GB of memory.

Figure 2.3 Titan supercomputer. *Courtesy of Oak Ridge National Laboratory.*

XK7 is a state-of-the-art machine that composed of both CPUs and GPUs. Specialized programming scheme required for such a hybrid system will be discussed in Chapter 8. XK7 is the second platform from Cray, Inc to employ a hybrid system. Besides Titan, Cray has also offered the product to the Swiss National Supercomputing Centre, with 272 node machines and Blue.

In the overall architecture of XK7, each blade contains four nodes, each with 1 CPU and 1 GPU per node. The system is scalable to 500 cabinets that can carry 24 blades.

Each CPU contains 16 cores of AMD Opteron 6200 Interlagos series and the GPU model is the Nvidia Tesla K20 Keplar series. Each CPU can be paired with 16 or 32 GB of error-correcting code memory, whereas GPUs can be paired with 5 or 6 GB of memory, depending on the model used. The interconnect is Gemini servicing two nodes with a capacity of 160 GB/s. Power consumption is between 45 and 54 kW for a full cabinet. These cabinets are air- or water-cooled. The OS is SUSE Linux, which can be programmed according to need.

3.3 Sequoia BlueGene/Q

Sequoia, an IBM BlueGene/Q system installed at DOE's Lawrence Livermore National Laboratory (LLNL), is again the number three system. It was first delivered in 2011 and achieved 17.17 PFLOP/s on the Linpack benchmark.

To achieve 17.7 PFLOP/s, Sequoia ran LINPACK for about 23 h with no core failure. However, division leader Kim Cupps, from the owner of the system, LLNL, said that the system is capable of hitting 20 PFLOP/s. In this way, it is more than 80% efficient, which is excellent. She added, "For a machine with 1.6 million cores to run for over 23 h 6 weeks after the last rack arrived on our floor is nothing short of amazing."

The system consumes 7890 kW of power. Cooling is basically done by running water through tiny copper pipes.

The cluster is extremely efficient for one so large, with 7,890 kW of power, compared with 12,659 kW for the second-best K computer. It is primarily cooled by water running through tiny copper pipes encircling the node cards. Each card holds 32 chips, each of which has 16 cores.

3.3.1 Sequoia's Architecture

The whole system is Linux-based. Computer node Linux runs on 98,000 nodes, whereas Red Hat Enterprise Linux runs on 786 input-output nodes that connect to the file system network.

Use of the Sequoia has been limited since February 2013 because, Kim Cupps said, "To start, the cluster is on a relatively open network, allowing many scientists to use it. But after IBM's debugging process is over around February 2013, the cluster will be moved to a classified network that isn't open to academics or outside organizations. At that point, it will be devoted almost exclusively to simulations aimed at extending the lifespan of nuclear weapons." She added that "The kind of science we need to do is lifetime extension programs for nuclear weapons, that requires suites of codes running. What we're able to do on this machine is to run large numbers of calculations simultaneously on the machine. You can turn many knobs in a short amount of time." Figure 2.4 shows the detailed architecture of the Sequoia supercomputer.

Blue Gene/Q uses a PowerPC architecture that includes hardware support for transactional memory, allowing more extensive real-world testing of technology.

In November 2011, in the Top500 list, only three of the top five supercomputers were gaining benefit from GPU. By 2013, more than 60 supercomputers were equipped with GPUs.

Devick Turek, IBM Vice President of High Performance Computing, said that "The use of GPUs in supercomputing tends to be experimental so far; the objective of this is to do real science."

3.4 K-Computer

Fujitsu's K-Computer, installed at the RIKEN Advanced Institute for Computational Science (AICS) in Kobe, Japan, is the number four system with 10.51 PFLOP/s on the Linpack benchmark. The K-supercomputer

Introduction to High-Performance Computing   29

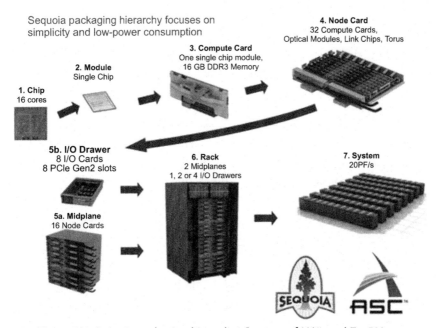

Figure 2.4 Sequoia packaging hierarchy. *Courtesy of LLNL and Top500.org.*

has been included in the top of the list in June 2011, Figure 2.5. The K-supercomputer is a Japanese supercomputer. The K-supercomputer has 68,544 CPUs, each with eight cores. This is twice as many as any other supercomputer on the Top500. The letter "K" is short for the Japanese word *Kei*, which means "Ten Quadrillions." Thus, the K-Computer is capable of performing 8 quadrillion calculations per second. It contains more than 800 racks. In 2004, the Earth Simulator was the world's fastest

Figure 2.5 K-computer. *Courtesy of Top500.org.*

computer (with a speed of about 36 teraFLOPs); in June 2011, Japan has held the first position for the first time in world ranking. Something strange thing about the K-Computer was that, like other top supercomputers, it contained neither NVIDIA cores nor the conventional x86 processors from Intel or AMD. It used Fujitsu-designed SPARC processors. The late Hans Warner Meuer (died in 2014) and Ex-Chairman of International Super Computers said, "The SPARC architecture breaks with a couple of trends which we have seen in the last couple of years in HPC and the Top500. It is a traditional architecture in the sense it does not use any accelerators. That's the difference compared to the number two and number four systems, which both use NVIDIA accelerators. The SPARC architecture breaks with a couple of trends which we have seen in the last couple of years in HPC and the Top500."

These sophisticated Japanese computers are also energy efficient. Although it is surprisingly power hungry and pulls over 9.8 MW of power to make its computations, the system is the second most power-efficient system in the entire Top500 list. In comparison, Rackspace's main UK data center consumes 3.3 MW. Hans Meuer told the audience at a conference, "It's not because of inefficiencies; it's simply driven by its size. It's very large computer." Based on developments, Meuer added that the ISC expected a system to reach an exaFLOP/s by around 2019. Intel General Manager of Data Centers, Kirk Skaugen said in a conference briefing, "It has impressive efficiency for sure, but it already uses up half the power at around 8 PFLOP/s of what we're going to try and deliver with 1,000 PFLOP/s [in 2018]." [2].

Another interesting feature of the K-Computer is the mesh-tours network to pass information between processes. The topology is designed so that each node is connected to six others, and individual computing jobs can be assigned to a number of interconnected processors. In addition, the Japanese machine uses an HPC six-dimensional mesh-torus network to pass information between processors. The network topology means that every computing node is connected to six others, and individual computing jobs can be assigned to groups of interconnected processors, at which point it will have around 15% more processors.

3.5 Mira, BlueGene/Q system

Mira, a BlueGene/Q system, is the fifth in the world, with 8.59 PFLOP/s on the LINPACK benchmark. The IBM Blue Gene/Q supercomputer, at the Argonne Leadership Computing Facility, is equipped with 786,432 cores and 768 terabytes of memory and has a peak performance of 10 PFLOP/s. Mira's 49,152 compute nodes have a PowerPC A2 1600-MHz processor containing 16 cores, each with four hardware threads, running at 1.6 GHz, and 16 GB of DDR3 memory. A 17th core is available for the communication library. The system nodes are connected in 5D torus topology and the hardware covered area is 1632 ft^2.

The system contains a quad Floating Point Unit (FPU) that can be used to execute scalar FPU. The Blue Gene/Q system also features a quad FPU that can be used to execute scalar floating point instructions, four-wide SIMD instructions, or two-wide complex arithmetic SIMD instructions. This quad FPU provides higher single-thread performance for some applications.

The file system is a general parallel file system that has a capacity of 24 PB and 240 GB/second bandwidth. All of these resources are available through high-performance networks including ESnet's recently upgraded 100-GB/s links [3]. The system is shown in Figure 2.6.

The performance of the world's five top supercomputers is listed in Table 2.1.

Since the advent of the IBM roadrunner, which touched the barrier of PFLOP/s in 2008, speed has become a yardstick to measure the performance of the world's supercomputers even until now. The next generation of these supercomputers has to cross the exa-scale line; all of the leading companies, such as HP, Dell, IBM, and Cray, are working hard to achieve massive exa-scale supercomputing.

Figure 2.6 The Mira supercomputer. *Courtesy of Argonne National Laboratory.*

Table 2.1 Peak performance of the top five supercomputers

	Tianhe 2	Titan, Cray XK 7	Sequoia	K-Computer	Mira, BlueGene
System family	Tianhe	Cray XK7	IBM BlueGene/Q	K-Computer	BlueGene
Installed at	NUDT	DOE's Oak Ridge National Laboratory	LLNL	AICS	Argonne Leadership Computing Facility
Application area	Research	Research	Research	Research	Research
Operating system	Linux (Kylin)	Linux	Linux	Linux	Linux
Peak performance	33.86 PFLOP/s	17.59 PFLOP/s	17.17 PFLOP/s	10.51 PFLOP/s	8.59 PFLOP/s

4. SOME BASIC BUT IMPORTANT TERMINOLOGY USED IN HPC

When a cluster is ready to run, a few steps are still needed to check its performance before particular applications are started. There are two steps in this regard: the first is the testing phase of the hardware and the second is benchmarking to make sure that the desired test requirements are met. For this purpose, several terms are used in HPC, so the reader must be familiar with them.

4.1 Linpack

Linpack is a code for solving matrices of linear equations and least-square problems. Linpack consists of FORTRAN subroutines and was conventionally used to measure the speed of supercomputers in FLOP/s. The term later overlapped with "linear algebra package" (LAPACK). LAPACK benchmarks are system floating point computing power. The scheme was introduced by Jack Dongarra and was used to solve an $N \times N$ system of linear equations of the form $Ax = b$. The solution was then obtained using the conventional Gaussian elimination method but with partial pivoting. Floating point operations are $2/3N^3 + 2N^2$, the results of which are presented in the form of FLOP/s. There have been some drawbacks to Linpack because it does not push the interconnect between nodes, but focuses on floating point arithmetic units and cache memory. Thom Dunning, Director of the National Center for Supercomputing Applications, said about Linpack: "The Linpack benchmark is one of those interesting phenomena—almost anyone who knows about it will decide its utility. They understand its limitations but it has mindshare because it's the one number we've all bought into over the years" [4].

The Linpack benchmark is a CPU- or compute-bound algorithm. Typical efficiency figures are between 60% and 90%. However, many scientific codes do not feature the dense linear solution kernel, so the performance of this Linpack benchmark does not indicate the performance of a typical code. A linear system solution through iterative methods, for instance, is much less efficient in an FLOP/s sense, because it is dominated by the bandwidth between CPU and memory (a bandwidth-bound algorithm).

4.2 LAPACK

The basic issue in CFD is storage of data. Data are usually in the form of linear equations that are compacted in matrix form. If you are an undergraduate, you have learned that linear equations can easily be written in matrix form and then solved either using iterative or direct methods. In a similar manner, a computer uses these methods to solve matrices.

4.2.1 Matrix Bandwidth and Sparse Matrix

In the case of implicit methods, the values of a variable at all points are computed at a new time step with a single value known at a previous time step. In this way, because there are many unknowns, the number of equations (which is only one), matrix is formed and solved with various procedures. If the approximation (the number of points taken to solve the difference equation) is three points, the matrix would be tri-diagonal. In the case of a five-point approximation, the matrix would be pent-diagonal. This means that the matrix would contain elements necessarily lying along diagonal to as well as one strip above and below it in the case of tri-diagonal, or two in the case of pent-diagonal; the rest of the elements in the matrix are zero. Such a matrix is called a sparse (widely open) matrix. In HPC, we take the advantage of this zero matrix operation.

The matrix form is shown in Figure 2.7. Here, the dominant elements lie in the diagonal stream. In terms of computing, what happens is that the diagonals are stored consecutively in memory. If the matrix is tri-diagonal with a three-point stencil, we can name the diagonal strip just from the first row, first column; then the top strip is a super-strip and the bottom strip is a sub-strip. The most economical storage scheme is $2n - 2$ elements consecutively.

The bandwidth of a matrix is defined as $b = k1 + k2 + 1$, where $k1$ is the left diagonal(s) below the main diagonal and $k2$ is the right diagonal(s) above the main diagonal that was mentioned as the sub-strip and super-strip

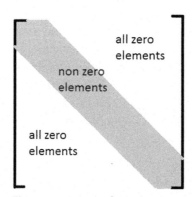

Figure 2.7 Image of sparse matrix.

just above. So if $k1 = k2 = 0$ then the matrix is diagonal only while if $k1 = k2 = 1$ then matrix is called tri-diagonal. However, if a matrix has dimensions n × n and bandwidth b, then for storage the matrix would be n × b. For example, consider the following matrix:

$$\begin{bmatrix} a_{11} & a_{12} & 0 & \cdots & \cdots & 0 \\ a_{21} & a_{22} & a_{23} & \ddots & \ddots & \vdots \\ 0 & a_{32} & a_{33} & a_{34} & \ddots & \vdots \\ \vdots & \ddots & a_{43} & a_{44} & a_{45} & 0 \\ \vdots & \ddots & \ddots & a_{54} & a_{55} & a_{56} \\ 0 & \cdots & \cdots & 0 & a_{65} & a_{66} \end{bmatrix}$$

In the case of diagonal storage, the computer shifts the first row toward the right, making the first element zero; i.e., for the above case, it would be a 6 × 3 matrix

$$\begin{bmatrix} 0 & a_{11} & a_{12} \\ a_{21} & a_{22} & a_{23} \\ a_{32} & a_{33} & a_{34} \\ a_{43} & a_{44} & a_{45} \\ a_{54} & a_{55} & a_{56} \\ a_{65} & a_{66} & 0 \end{bmatrix}$$

where the first and last elements represent the zero upper right and lower left triangles, respectively. For conversion between array elements A(i,j) and matrix elements A_{ij}, this can be easily done in Fortran. If we allocate the array with dimension

```
dimension A(n,-1:1)
```

the main diagonal A_{ii} is stored in A(*,0). For instance, A(2,0) is similar to a_{21}. The next location in the same row of matrix A would be A(2,1), which is similar to a_{22}. We can write this in a general format as

```
A(1,j) ~ a_{i,i+j}
```

4.2.2 Lower Upper (LU) Factorization

As mentioned earlier, LAPACK is the linear algebra package. The package contains subroutines for solving systems of simultaneous linear equations, least-square solutions of linear systems of equations, eigenvalue, and singular value problems. LAPACK software for linear algebra works on an LU factorization routine that overwrites the input matrix with the factors.

The original goal of LAPACK was for Linpack libraries to run efficiently on shared memory vector and parallel processors. LAPACK routines are written so that, as much as possible, computations are performed by calling up a library called Basic Linear Algebra Sub-programs (BLAS). The BLAS helps LAPACK to achieve high performance with the aid of portable software.

Certain levels of operations are built into these libraries: levels 1, 2, and 3. Level 1 performs vector operations. Level 2 performs matrix–vector operations such as matrix vector product and implicit calculations such as LU decomposition. Level 3 defines matrix–matrix operations, most notably the matrix–matrix product. LAPACK uses the blocked operations of BLAS level 3. This is surely to achieve high performance on cache-based CPUs. However, with modern supercomputers, several projects have come up that extended LAPACK functionality to distributed computing, such as Scala-pack and PLapack (Parallel LAPACK) [5].

4.2.3 LU Factorization of Sparse Matrices
Consider a tri-diagonal matrix, as shown below. Applying Gaussian elimination will lead to a second row in which one element is eliminated. One important point is that the zero elements did not become disturbed owing to Gaussian operation. Second, the operation did not affect the tri-diagonality of the matrix. The treatment suggests that L + U factorization takes the same amount of memory as the previous method; however, the case is not limited to tri-diagonal matrices.

$$
\begin{bmatrix}
2 & -1 & 0 & \dots \\
-1 & 2 & -1 & \\
0 & -1 & 2 & -1 \\
 & \ddots & \ddots & \ddots & \ddots
\end{bmatrix}
\Rightarrow
\left[
\begin{array}{c|cccc}
2 & -1 & 0 & \dots \\
\hline
0 & 2-1/2 & -1 & \\
0 & -1 & 2 & -1 \\
 & \ddots & \ddots & \ddots & \ddots
\end{array}
\right]
$$

4.2.4 Basic Linear Algebra Sub-program Matrix Storage
Following are key factors on the basis of which matrices are stored in BLAS and LAPACK.

4.2.4.1 Array Indexing
In a FORTRAN environment, 1-based indexing is used, whereas C/C++ routines usually use index arrays such pivot points in LU factorization.

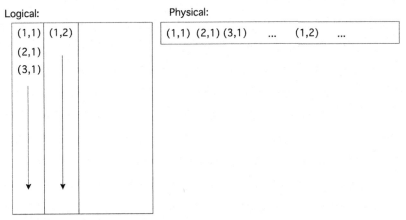

Figure 2.8 Storage of an array in FORTRAN with respect to column major.

4.2.4.2 FORTRAN Column Major Ordering

FORTRAN stores the indices in the form of column priority, i.e., elements in a column are stored consecutively. Informally, programmers call thus this trick "the leftmost varies quickest." See Figure 2.8 for how the elements are stored.

4.2.4.3 Submatrices and Leading Dimension of A Parameter

Using the above storage scheme, it is evident how to store an $m \times n$ matrix in mn memory locations. But, there are many cases in which software needs access to a matrix that is a sub-block of another, larger matrix. The sub-block is no longer being shared in memory, as can be seen in Figure 2.9. The way to describe this is to introduce a third parameter in addition to M, N: the leading dimension of A, or the allocated first dimension of the surrounding array, which is illustrated in Figure 2.10.

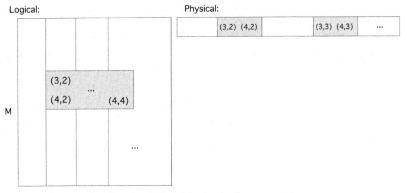

Figure 2.9 Sub-block of a larger matrix.

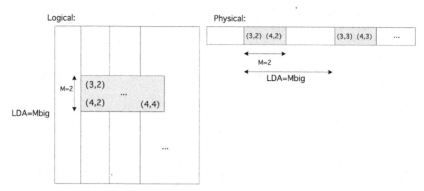

Figure 2.10 Sub-block of a larger matrix but using leading dimension of LDA.

Now, suppose that you have a matrix A of size 100×100 which is stored in an array 100×100. In this case LDA is the same as m. Now suppose that you want to work only on the submatrix A(91:100 , 1:100); in this case the number of rows is 10 but LDA=100. Assuming the fortran column–major ordering (which is the case in LAPACK), the LDA is used to define the distance in memory between elements of two consecutive columns which have the same row index. If you call B = A(91:100 , 1:100) then B(1,1) and B(1,2) are 100 memory locations far from each other according to the following formula, B(i,j)=i+(j−1)*LDA, implies that, B(1,2)=1+(2−1) *100= 101th location in memory.

4.2.5 Organization of Routines in LAPACK

LAPACK is organized through three levels of routines:

1. Drivers. These are powerful, top-level routines for problems such as solving linear systems or computing a singular value decomposition (SVD). In linear algebra, the SVD is a decomposition of a real or complex matrix. It is mostly used in scientific computing and statistics. Formally, the singular value decomposition of an m × n real or complex matrix **M** is a factorization of the form $\mathbf{M} = \mathbf{U\Sigma V^\star}$, where **U** is an m × m real or complex unitary matrix, meaning that $\mathbf{UU^\star} = \mathbf{I}$, $\mathbf{\Sigma}$ is an m × n rectangular diagonal matrix with non–negative real numbers on the diagonal, and $\mathbf{V^\star}$ (the conjugate transpose of **V**, or simply the transpose of **V** if **V** is real) is an n × n real or complex unitary matrix. The diagonal entries $\mathbf{\Sigma}_{i,i}$ of $\mathbf{\Sigma}$ are known as the singular values of **M**. The m columns of **U** are called the left-singular vectors and the n columns of **V** are called the right-singular vectors of **M**.

2. Computational routines. These are the routines of which drivers are built. A user may have occasion to call them by themselves.

3. Auxiliary routines:

Routines conform to a general naming scheme: XYYZZZ where,

X is the precision: S, D, C, Z stands for single and double, single complex and double complex, respectively.

YY is the storage scheme: general rectangular, triangular, and banded. It indicates the type of matrix (or the most significant matrix). Most of these two-letter codes apply to both real and complex matrices; a few apply specifically to one or the other.

BD: Bi-diagonal.

DI: Diagonal.

GB: General band.

GE: General (i.e., asymmetric or rectangular).

GG: General matrices, generalized problem (i.e., a pair of general matrices).

GT: General tri-diagonal.

HB: (Complex) Hermitian band.

HE: (Complex) Hermitian.

HG: Upper Hessenberg matrix, generalized problem (i.e., a Hessenberg and a triangular matrix).

HP: (Complex) Hermitian, packed storage.

HS: Upper Hessenberg.

OP: (Real) orthogonal, packed storage.

OR: (Real) orthogonal.

PB: Symmetric or Hermitian positive definite band.

PO: Symmetric or Hermitian positive definite.

PP: Symmetric or Hermitian positive definite, packed storage.

PT: Symmetric or Hermitian positive definite tri-diagonal.

SB: (Real) symmetric band.

SP: Symmetric, packed storage.

ST: (Real) symmetric tri-diagonal.

SY: Symmetric.

TB: Triangular band.

TG: Triangular matrices, generalized problem (i.e., a pair of triangular matrices).

TP: Triangular, packed storage.

TR: Triangular (or in some cases quasi-triangular).

TZ: Trapezoidal.

UN: (Complex) unitary.

UP: (Complex) unitary, packed storage.

ZZZ is the matrix operation. They indicate the type of computation performed. For example, SGEBRD is a single precision routine that performs a bi-diagonal reduction BRD of a real general matrix. If it is in qrf form, the factorization is QR factorization, and for lqf it is LQ factorization. Thus, the routine sgeqrf forms the QR factorization of general real matrices in single precision; consequently, the corresponding routine for complex matrices is cgeqrf.

The names of the LAPACK computational and driver routines for the FORTRAN 95 interface in an Intel compiler are the same as FORTRAN 77 names but without the first letter that indicates the data type. For example, the name of the routine that forms the QR factorization of general real matrices in the FORTRAN 95 interface is geqrf. Handling different data types is done by defining a specific internal parameter referring to a module block with named constants for single and double precision.

REFERENCES

[1] 42nd edition of the twice-yearly Top500 list of the world's most powerful super-computers, November 18, 2013 at the SC13 Conference, Denver, Colorado.
[2] Inside Japan's Top500 K-Computer. Cited July 11, 2011, http://www.zdnet.co.uk/news/emerging-tech/2011/06/20/japan-takes-top500-supercomputer-crown-from-china-40093151/, article by Jack Clark | June 20, 2011 – 11:38 GMT (04:38 PDT), (accessed September 2011).
[3] https://www.alcf.anl.gov/user-guides/mira-cetus-vesta, (accessed 15.06.14.).
[4] Mims Christopher. Why China's New Supercomputer Is Only Technically the World's Fastest. Website. 2010. Cited September 22, 2011.
[5] Van de Geijn Robert A. Using PLAPACK: Parallel Linear Algebra Package. The MIT Press; 1997.

CHAPTER 3

The Way the HPC Works in CFD

1. INTRODUCTION

As mentioned earlier, computational fluid dynamic (CFD) applications are resource hungry. Their major areas of application are aeronautics and propulsion. It is now of paramount importance to perform high-fidelity simulations because experiments within these two domains are costly. Thus, high performance is the best solution: one that that gives flow results in a reasonable amount of time. One might ask which systems are classified as high-performance computers (HPC). There is no one exact answer, but from a current technology perspective and the latest computational speeds, all systems that have computational power of more than 10^{12} floating points per second (FLOPS/s) are HPC. In HPC, two kinds of architecture are common. The first is vector, which has high memory bandwidth and vector processing capabilities. The second architecture relies on a cache-based memory interconnect to a shared memory system (symmetric multiprocessing); because it is cost-effective, this type of system has rapidly captured the market. This indication has been prompted by the Peak Performance Gordon Bell Prize, which is awarded each year at the International Supercomputing Conference. It shows that scalar platforms have dominated over vector supercomputers for the past 3 consecutive years. According to Keyes et al. [1], a difficulty observed with cached-based processors is that computing efficiency is less than 20% for most CFD applications. The reason is that because flow solvers process a large amount of data for CFD (if we consider large eddy or direct numerical simulations), a lot of communication is required within the processor itself.

With scalar-based central processing units (CPUs), there is a bit issue for most CFD applications. As mentioned by Keyes [1], computing efficiency is less than 20%. This is because the communication required by the processor within different cache levels needs to handle a lot of data owing to the massive information of CFD calculations. For a mesh size of 5 million mesh size and six equations (variables) and if it is three-dimensional (3D), that would make it 5 million × 6 × 3 = 90 million cell information for just one

Using HPC for Computational Fluid Dynamics
ISBN 978-0-12-801567-4
41

iteration. Such tasks need to run on a large number of computing cores and parallel platforms have made it available to access large amount of memory bank. They integrate all of the processor cores (a processing unit (PU) corresponds to the task performed by one computing core). Programming such a parallel solver becomes difficult when there must be data sharing between processors as well apart from solving the problem. If parallel cores are less than 128, efficiency relies on communication between processor cores. If a large number of cores are used, there is load balancing in addition to communication among the CPU cache; this is the limiting factor of efficiency.

2. OLDEST FORM OF PARALLEL COMPUTING

Vector computing was used for initial operations on supercomputers, such as matrix addition, which could be performed on independent sub-blocks. Vector computing implied vector operations in the instruction set. In this form of parallel architecture, two different types were used:

1. Multiple functional units (in which the CPU was able to perform multiple operations simultaneously)
2. Pipelining

A number of assembler statements were required to perform even simple operations. It performed one assembler operation at a time, rather than one single mathematical operation.

Vector computing is good if the algorithms can be formulated in terms of vector–vector operations. A practical example is the Earth Simulator located at NEC, in Japan.

3. HIGH-PERFORMANCE COMPUTING SOLVER PLATFORM

Modern flow solvers have to run efficiently on a large range of super-computers, from single core to massively parallel platforms. The most common system used for HPC applications is Michael Flynn taxonomy, which divides HPC programming into two classed, single instruction multiple data (SIMD) and multiple instruction multiple data (MIMD)-based platforms. It can be further divided in two subcategories:

1. SIMD: In this case, multiple individual processors (cores) simultaneously execute the same program (but at independent points) on different data. In SIMD, each processor performs the same arithmetic operation (or stays idle) during each computer clock cycle, and is controlled by a single

central control unit. This is shown in Figure 3.2. Communication and computation are synchronized precisely at every clock period.

2. MIMD: Multiple autonomous processors simultaneously operate at least two independent programs. Typically, such systems pick one processor to be the host (the master), which runs one program that distributes data to all other processors (the slaves), all of which run a second program. As in ANSYS FLUENT and CFX, the mesh is split (by different techniques) among various cores by the master node.

3.1 Parallel Efficiency

A famous strategy is "divide and rule." Parallel efficiency deals with this concept. It is easier to solve a problem by splitting it into pieces rather than by grasping the whole bite. In CFD, the strategy is to break up the original problem into a set of tasks. Then, as said earlier, MIMD tasking starts and is used to solve the problem. Usually, the task is divided into N subtasks and each task is solved by one CPU. In ANSYS FLUENT, the mesh is divided into as many partitions as the number of cores (by default), so each core solves one partition. The user may increase the number of partition solved by each block. If N cores are working in parallel, the time for sequential or single cores to run the same problem must be N multiplied by the time taken by each PU. This is not usually achieved because of several bottlenecks such as CPU cache communication, memory bandwidth, and network latency. One important aspect in this regard is the reliability of the parallel system. This is discussed in the next section on scalability.

3.2 Task Scalability

Scalability measures whether a given problem can be solved more quickly as more processors are used. Speedup S is defined as the ratio between the time taken to run a job in parallel, $T_{parallel}$, and the time to run a job in series, $T_{sequential}$, i.e.,

$$S = \frac{T_{parallel}}{T_{Sequential}} \qquad (3.1)$$

It is used to quantify the scalability of the problem. Typically, a value of S equal to the number of available PUs corresponds to fully parallel tasks. Computing efficiency E corresponds to the value of S divided by the number of computing cores N, as discussed above. Ideally, then, $S = N$ so $E = 1$. Thus, if efficiency is poor, that means the problem is not correctly balanced among all PUs (load balancing error) or that the time needed for communication is important compared with computing time.

Here, we divert to the architecture of computer speeds, which is the basic root of scalability. From this, we will discuss Amdahl's law.

3.2.1 Clock Speeds

Normally, the speed that matters a lot is the clock speed. This can be understood by a simple example. Consider a program that takes 12 s to run on computer A and has a 400-MHz clock. We are trying to build a newer machine, B, which will run this program in 8 s. This new design will substantially increase the clock rate, thereby affecting the rest of the CPU design, causing machine B to require 1.2 times as many clock cycles as machine A. What clock rate should we propose?

Pretty interesting! We can now calculate the cycles of machine A with the method:

Clock cycles for A $= 12$ s $\times$ 400 M cycles/s $= 4.8 \times 10^9$ cycles.

Now, the clock cycles (C.C) for B will be $1.2 \times$ C.C A $= 1.2 \times 4.8 \times 10^9$ cycles $= 5.76 \times 10^9$ cycles.

Thus, the clock rate for B $=$ clock cycles/execution time of B $= 5.76 \times 10^9/6$ s $= 960$ MHz.

3.2.2 Cycles per Instruction

Another important parameter after clock speed is the Cycles Per Instruction (CPI). On the basis of CPI, it is not straightforward that a machine with a higher CPI would be faster. We will need to check the time per instruction. For example, machine A has a clock time of 10 ns with a CPI of 2.0, and B has a clock time of 20 ns with a CPI of 1.2. Then,

Time per instruction for A $= 10$ ns $\times$ 2 CPI $= 20$ ns.

Time per instruction for B $= 20$ ns $\times$ 1.2 CPI $= 24$ ns.

Thus, A is $24/20 = 1.2$ times faster than B. It is a coincidence that the machine with higher CPI won, but it also depends on the clock time, which should be taken care of.

Another scenario is with some sort of sequence of multiple operations in a program. Let us assume a situation in which there are two options to run three different classes of instruction: A, B, and C, requiring one, two, and three cycles, respectively.

Let us say that the first sequence has five instructions: two of A, one of B, and two of C.

The second sequence has six instructions: four of A, one of B, and one of C. We now want to check which sequence of operations will be faster. We also want to calculate the CPI.

Sequence 1: $2 \times 1 + 1 \times 2 + 2 \times 3 = 10$ cycles; $CPI_1 = 10/(2 + 1 + 2) = 2$

Sequence 2: $4 \times 1 + 1 \times 2 + 1 \times 3 = 9$ cycles; $CPI_2 = 9/(4 + 1 + 1) = 1.5$ Thus, sequence 2 is faster.

3.2.3 Million Instructions per Second

Another important parameter that is used to measure performance is Millions of instructions per second (MIPS). If we divide the instruction count by the execution time and express it as 1×10^6, we get MIPS. Millions of instructions per second depend on clock frequency and CPI. It varies inversely with the performance. Let us see an example:

Example: Two compilers are being tested for a 200-MHz machine with three different classes of instructions: A, B, and C, which require one, two, and three cycles, respectively. Compiler 1: Compiled code uses 5 million class A, 1 million class B, and 1 million class C instructions. Compiler 2: Compiled code uses 10 million class A, 1 million class B, and 1 million class C instructions.

1. Which sequence will be faster according to MIPS?

2. Which sequence will be faster according to execution time?

First, we calculate the cycles and instructions.

For compiler 1: 10 million cycles through $(5 \times 1 + 1 \times 2 + 1 \times 3)$ and 7 million instructions $(5 + 1 + 1)$

For compiler 2: 15 million cycles through $(10 \times 1 + 1 \times 2 + 1 \times 3)$ and 12 million instructions $(10 + 1 + 1)$

We know that execution time = clock cycles/clock rate (cycles per second)

Therefore, for compiler 1, $E_1 = 10 \times 10^6/100 \times 10^6 = 0.1$ s.

For compiler 2, $E_2 = 15 \times 10^6/100 \times 10^6 = 0.15$ s.

It is time to computer MIPS = instruction count in millions/(execution time $\times 10^6$)

Thus, $MIPS_1 = 7 \times 10^6/0.1 \times 10^6 = 70$ and $MIPS_2 = 12 \times 10^6/ 0.15 \times 10^6 = 80$.

It can be concluded that compiler 2 uses more single-cycle instructions whereas compiler 1 is faster with respect to the execution time.

We have seen the three aspects of evaluating performance for parallel computing on the basis of cycles per second or the clock rate; the second one depends on the number of cycles per instruction and MIPS. These calculations are important when designing a code (not only for CFD but) to perform efficiently on parallel cores.

For overall job distribution on cores, a criterion is specified, which is discussed in the next section.

4. AMDAHL'S LAW

Amdahl's law states that speedup can be defined in terms of part of the code (let's say P) that can be parallelized, so

$$\text{Speedup} = 1/(1 - P) \tag{3.2}$$

If no part of a code can be parallelized, then $P = 0$ and the speedup $= 1$ (no speedup). Conversely, if all of the code can be parallelized, $P = 1$ and the speedup is infinite (in theory). Similarly, if 50% of the code can be parallelized, maximum speedup $= 2$, meaning that the code will run twice as fast. In computing terms, this can be defined with regard to the number of processors performing the parallel fraction of work, so the above equation can be modified as:

$$S = 1/[(P/N) + S] \tag{3.3}$$

where $P =$ parallel fraction, $N =$ number of processors (or cores), and $S =$ serial fraction. This means that to achieve good speedup we must have a greater number of cores that perform parallel fraction faster. But if the problem is not big enough to divide its portions into a certain number of cores, this speed would start to drop. This means that much time is consumed in communicating data between cores. A parameter called granularity, which is the ratio of computation over communication, decreases. Ideally, if these kinds of overheads are skipped, then upon increasing cores the speedup will always tend to increase.

Certain problems, however, occur with increased performance by increasing the problem size. For example, consider a problem with 2D grid calculations as the parallel portion.

| 2D Grid calculations | 85 s | 85% |
| Serial fraction | 15 s | 15% |

We can increase the problem size by doubling the grid dimensions and halving the time step. This results in four times the number of grid points and twice the number of time steps. The timings then look like:

2D Grid calculations	680 s	97.84%
Serial fraction	15 s	2.16%

We now know that problems that increase the percentage of parallel time with their size are more scalable than problems with a fixed percentage of parallel time.

Consider another simple example:

If it were possible to parallelize 90% of a program, the remaining 10% of code would run sequentially; then, according to Amdahl's law:

$T_{parallel} = (0.9 \times T_{serial})/P + 0.1 \times T_{serial}$ where P = the number of available cores (processors).

If $T_{serial} = 20$ s and P = 6, then speedup will be: $S = 20/(18/P + 2) = 4$.

The time spent in the parallel portion of code decreases as the number of cores increases. Eventually this time tends toward zero but the time spent in the sequential part of the code still remains and strongly limits program speedup. Here, we assumed a constant problem size and only changed the number of cores, irrespective of the above example.

As a consequence, Amdahl's law says that speedup will always be less than 1/r, where r is the sequential portion of the program.

Amdahl's law assumes that all processes finish their task at the same time and one process is then dedicated to a sequential task $(1 - P)$. This assumption of perfect load balancing is not true in most CFD applications, especially when an industrial configuration is considered. In fact, each process ends its task at a different time and the value of $(1 - P)$ is no longer meaningful. To deal with this problem, Amdahl's law can be extended by taking into account the working time of each process [2]. A typical scheme of a parallel computation is shown in Figure 3.1, with the time spent by each process to perform its task.

By convention, the overall time related to the parallel computation $T_{parallel}$ is set to 1. The sequential computation time $T_{sequential}$ is thus estimated by adding the time spent by each process, and is expressed by [2]:

$$T_{speedup} = NP_0 + (N - 1)P_1 + \cdots + 2P_{N-2} + P_{N-1} \qquad (3.4)$$

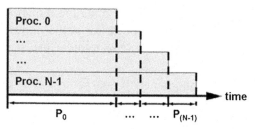

Figure 3.1 Time taken by processors.

where P_{i+1} is the part of the work that is computed with $(N - i)$ computing cores sum of all P_i where $i = 0$ to $N - 1$ is equal to unity. Then speedup can automatically be found as the ratio of $T_{sequential}$ to $T_{parallel}$.

$$Speedup = \sum_{i=0}^{N-1} (N - i)P_i \qquad (3.5)$$

5. DATA HANDLING AND STORAGE

Two kinds of data need to be stored in an HPC environment:
1. Array or matrices
   Usually stored as either
   a. Block distributions
   b. Cyclic and block cyclic distributions
2. Irregular data
   An example could be data associated with an unstructured mesh; they can be handled with graph partitioning.
   The big task is to handle the matrices that need to be mapped onto an array of processors. For simplicity here, we are explaining the 1D and 2D decomposition case for mesh partitioning.

5.1 Gaussian Elimination: Method for Dense Matrices

In a Gaussian elimination procedure, one first needs to find a pivot element in the set of equations. This element is then used to multiply (or divide or subtract) the various elements from other rows to create zeros in the lower left triangular region of the coefficient matrix. Our main concern is that somehow the coefficient matrix will be reduced so that the last element of the matrix will give the correct value of the last variable directly by dividing it by the last element of matrix B, which is usually a matrix of constants. Throughout the procedure, we perform elementary row operations. We create zeros in the left triangular matrix to make it sparse. This means that most elements of the matrix are zero. To perform Gaussian elimination, for a single row operation, a computer needs to perform several steps. It needs to multiply the second row with a factor including the pivot element as well as the first coefficient of the second row, and then to subtract the second row from the first. If this has to be performed for a million equations, the process becomes lengthy for a computer as well.

An algorithm for Gaussian elimination is mentioned below:

```
1. Forward Elimination
     Do k = 1,N – 1              %(increment stage counter)
         Do i = k + 1,N          %(increment row index)
             mᵢₖ = aᵢₖ/aₖₖ       %(calculate Gauss multiplier)
             bᵢ = bᵢ – mᵢₖbₖ     %(modify RHS of ith equation)
             Do j = k + 1,N      %(increment column index)
                 aᵢⱼ = aᵢⱼ – mᵢₖaₖⱼ   %(modify matrix components of
         ith equation)
     Repeat j
   Repeat i
 Repeat k

%2. Backward Substitution
             xₙ = bₙ/aₙₙ
     Do i = N – 1, 1
         xᵢ = 0
         Do j = i + 1,N
             xᵢ = xᵢ + aᵢⱼxⱼ
         Repeat j
         xᵢ = (bᵢ – xᵢ)/aᵢᵢ
     Repeat i
```

This algorithm shows that we need $O(N^3)$ arithmetic operations to obtain a solution to the system of linear equations using Gaussian elimination. Each loop runs $O(N^3)$ times. The backward substitution $O(N^3)$ operation is negligible compared with $O(N^3)$ when considering a larger system of equations. In modern supercomputers, there is a dire need to improve the operation count.

5.2 One-Dimensional Domain Decomposition

One-dimensional domain decomposition is also called stripped mapping. This mapping in matrix form could be either row-wise (the processors have horizontal stripes) or column-wise (the processors have vertical stripes) along with three possible indexing schemes (block, cyclic, and block–cyclic). In the 1D case np rows are assigned to a processor.

5.3 Two-Dimensional Domain Decomposition

In 2D domain decomposition (or checkerboard mapping), for 2D distribution, an $n\sqrt{P} \times n\sqrt{P}$ size block is assigned to a process. It extends 1D domain decomposition with division of the matrix in two directions (vertical and horizontal). Again, block and block–cyclic mapping are possible.

Consider an $n \times n$ dense matrix multiplication. Let $C = A \times B$ using p processors; then, for decomposition based on the output data, each entry of C uses the same amount of computation. For 1D, np rows are assigned to a processor and for 2D, an $n\sqrt{P} \times n\sqrt{P}$ size block is assigned to a processor. Multidimensional distribution allows a higher degree of concurrency and can also help reduce interactions. Figure 3.2 explains the scheme.

Another diagram identifies the theory of decomposition in Figure 3.3.

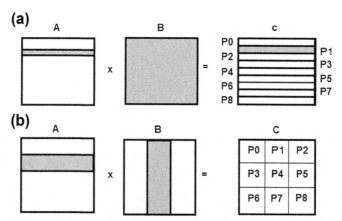

Figure 3.2 Data sharing needed for matrix multiplication with (a) 1D partitioning of the output matrix. (b) 2D partitioning of the output matrix. Shaded portions of the input matrices A and B are required by the process that computes the shaded portion of output matrix C.

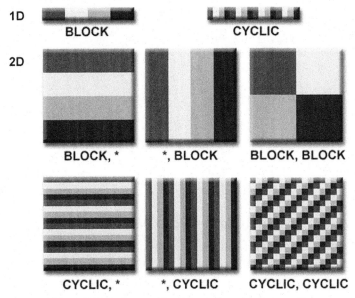

Figure 3.3 Decomposition types shown in grayscale.

5.4 Doolittle's Decomposition

$$
\begin{bmatrix}
a_{11} & a_{12} & \cdots & a_{1n} \\
a_{21} & a_{22} & \cdots & a_{2n} \\
\vdots & \vdots & \ddots & \vdots \\
a_{n1} & a_{n2} & \cdots & a_{nn}
\end{bmatrix} = LU
$$

$$
= \begin{bmatrix}
1 & 0 & \cdots & 0 \\
l_{21} & 1 & \cdots & 0 \\
\vdots & \vdots & \ddots & \vdots \\
l_{n1} & l_{n2} & \cdots & 1
\end{bmatrix}
\begin{bmatrix}
u_{11} & u_{12} & \cdots & u_{1n} \\
0 & u_{22} & \cdots & u_{2n} \\
\vdots & \vdots & \ddots & \vdots \\
0 & 0 & \cdots & u_{nn}
\end{bmatrix}
$$

By matrix–matrix multiplication.

$$u_{1j} = a_{1j} \quad j = 1, 2, \ldots, n \ (\text{1st row of U})$$

$$l_{j1} = a_{j1} / u_{11} \quad j = 1, 2, \ldots, n \ (\text{1st column of L})$$

5.4.1 Programming Structure

A Fortran-based simple code for Doolittle's decomposition is given below:

```
subroutine COL_LU(A)
!Manipulating L matrix
```

```
do k = 1,n
  do j = k,n
    A[j,k] = A[j,k]/A[k,k]
  enddo
  do j = k + 1,n
    do i = k + 1,n
      A[i,j] = A[i,j] − A[i,k]*A[k,j]
    Enddo
!!!!After this iteration, column A[k + 1:n,k] is logically the kth
column of L and row A[k,k:n] is logically the kth row of U.
  Enddo
Enddo
End
```

5.5 Two-Dimensional Block–Cyclic Distribution

For convenience we will number the processors from 0 to P − 1, and matrix columns (or rows) from 1 to N. In all cases, each sub-matrix is labeled with the number of the process (from 0 to 3) that contains it. Processor 0 owns the shaded sub-matrices, as shown in Figure 3.4.

Consider the layout illustrated on the left of Figure 3.2, 1D block column distribution. This distribution assigns a block of contiguous columns of a matrix to successive processes. Each process receives only one block of columns of the matrix. Column k is stored on process k/t_c, where $t_c = N/P$ is the maximum number of columns stored per processor. In the figure, N = 16 and P = 4. This layout does not permit good load balancing for the above Gaussian elimination algorithm because as soon as the first two columns are complete, process 0 is idle for the rest of the computation. The transpose of this layout, 1D block row distribution, has a similar loss for dense computations.

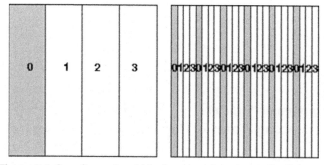

Figure 3.4 One-dimensional block and cyclic column distributions.

The second layout illustrated on the right of Figure 3.2, 1D cyclic column distribution, addressed this problem by assigning column k to process (k − 1) mod P. In the figure, N = 16 and P = 4. With this layout, each process owns approximately 1/Pth of the square southeast corner of the matrix, so the load balance is good. However, one shortcoming is that because single columns (rather than blocks) are stored, we cannot use the level 2 Basic Linear Algebra Subroutine (BLAS) (matrix–vector multiplication) to factorize and may not be able to use the level 3 BLAS (matrix–matrix multiplication) to update. The transpose of this layout, 1D cyclic row distribution, has a similar disadvantage.

The third layout shown on the left of Figure 3.5, the 1D block–cyclic column distribution, is a compromise between the distribution schemes shown in Figure 3.4. We choose a block size, NB, divide the columns into groups of size NB, and distribute these groups in a cyclic manner. This means column k is stored in processor $[(k − 1)/NB]|P|$. In this layout, $NB = t_c = [N/P]$ and $NB = 1$. In the figure $N = 16$, $P = 4$ and $NB = 2$. For NB larger than 1, this has a slightly worse balance than 1D cyclic column distribution but it can use level 2 BLAS and level 3 BLAS for local computations. For NB less than t_c, it has a better load balance than the 1D block column distribution but it can call up the BLAS only on smaller sub-problems. Hence, it takes less advantage of the local memory hierarchy. Moreover, this layout has the disadvantage that factorization will take place on one process, which represents a serial bottleneck.

This serial bottleneck is relieved by the fourth layout shown on the right of Figure 3.5, the 2D block cyclic distribution. Here, we think of our P processes arranged in a $P_r \times P_c$ rectangular array of processes, indexed in a 2D fashion by (p_r, p_c), with $0 < p_r < P_r$ and $0 < p_c < P_c$. All of the

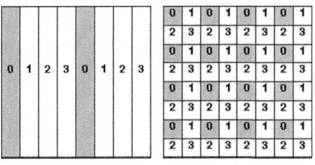

Figure 3.5 The 1D block–cyclic column and 2D block-cyclic distributions.

processes (p_r, p_c) with a fixed p_c are referred to as processor column p_c. All of the processes (p_r, p_c) with a fixed p_r are referred to as process row p_r. Thus, this layout includes all previous layouts and their transposes as special cases. In the figure, $N = 16$, $P = 4$, and $MB = NB = 2$. This layout permits P_c-fold parallelism in any column, and calls up level 2 BLAS and level 3 BLAS on local sub-arrays. Finally, this layout features good scalability properties. The 2D block–cyclic distribution scheme is the data layout used in the ScaLAPACK library for dense matrix computations.

6. TREATMENT OF CFD ALGORITHMS FOR PARALLEL PROCESSING

If parallelization is performed at the loop level (as is the case with auto-parallelizing compilers), Amdahl's law, which essentially says that the speed is determined by the least efficient part of the code, comes into play. To achieve high efficiency, the portion of the code that cannot be parallelized has to be very small. The best approach is to divide and rule: that is, to subdivide the solution domain into sub-domains and assign each sub-domain to one processor. In this case, the same code runs on all processors, on its own set of data. Because each processor needs data that reside in other sub-domains, the exchange of data between processors and/ or storage overlap is necessary. **Explicit schemes** are relatively easy to parallelize because all operations are performed on data from preceding time steps. It is necessary to exchange the data only at the interface regions between neighboring sub-domains after each step is completed. The sequence of operations and the results are identical on one and many processors. **Implicit methods**, on the other hand, are more difficult to parallelize. Calculation of the coefficient matrix and the source vector uses only old data and can be efficiently performed in parallel; the solution of linear equation systems is not easy to parallelize. For example, Gauss elimination, in which each computation requires the result of the previous one, is difficult to perform on parallel machines. Some other solvers can be parallelized and perform the same sequence of operations on n processors as on a single one, but either they are not efficient or the communication overhead is large. Basically, two schemes or approaches can be applied here.

The red–black Gauss–Seidel method is well-suited for parallel process-ing. It is often used in conjunction with multi-grid methods. On a struc-tured grid, the nodes are imagined to be colored in the same way as a checkerboard. The method consists of two Jacobi steps: black nodes are updated first, and then the red nodes. When the values at black nodes are updated, only the old red values are used. On the next step, red values are recalculated using the updated black values. This alternate application of the Jacobi method to the two sets of nodes gives an overall method with the same convergence properties as the Gauss–Seidel method. The nice feature about the red–black Gauss–Seidel solver is that it both vectorizes and parallelizes well, because there are no data dependencies in either step. In 2D, the nodes are colored as on a checkerboard; thus, for a five-point computational set, Jacobi iteration applied to a red point calculates the new value using data only from black neighbor nodes, and vice versa. The convergence properties of this solver are exactly those of the Gauss–Seidel method, which gave the method its name.

Computation of new values on either set of nodes can be performed in parallel; all that is needed is the result of the previous step. The result is the same as on a single processor. Communication between processors working on neighbor sub-domains takes place twice per iteration, after each set of data is updated. This local communication can be overlapped with computation of the new values. This solver is suitable only when used in conjunction with a multi-grid method, because it is inefficient.

ILU-type methods are recursive, which makes parallelization less straightforward. In this method, the elements of the L and U matrices depend on the elements at the left and bottom nodes. We call them W and S nodes, respectively. One cannot start calculating the coefficients on a sub-domain, other than the one in the southwest corner, before data are obtained from its neighbors. In 2D, the best strategy is to subdivide the domain into vertical stripes, i.e., use a 1D processor topology. Computation of L and U matrices and iteration can then be performed fairly efficiently in parallel c.f [3] earlier referred by Bastian [4]. The processor for sub-domain 1 needs no data from other processors and can start immediately; it proceeds along its bottom or lowest line. After it has calculated the elements for the rightmost node, it can pass those values to the processor for sub-domain 2. While the first processor starts calculation on its next line, the second one can compute on its bottom line. All n processors are busy when the first one reaches the nth line from bottom. When the first processor reaches the top boundary, it has to wait until the last processor—which is n lines behind—finishes the calculations (Figure 3.6).

In the iteration scheme, two passes are needed. The first is done in the manner just described whereas the second is essentially its mirror image.

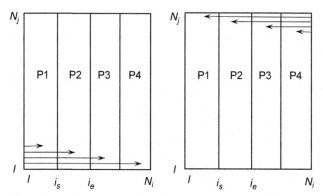

Figure 3.6 Parallel processing in the ILU solver in the forward loops (left) and in the backward loops (right).

A possible algorithm can be written as mentioned below. Please note that subscripts E, W, S, and N denote east, west, south, and north, respectively.

The algorithm is as follows:

```
for j = 2 to Nⱼ − 1 do:
  receive Uₑ(iₛ − 1,j),Uₛ(iₛ − 1,j) from west neighbor;
  for i = iₛ to iₑ do:
    calculate Uₑ(i,j),Lᵥ(i,j),Uₙ(i,j),Lₛ(i,j),Lₚ(i,j);
  end i;
  send Uₑ(iₑ,j), Uₙ(iₑ,j) to east neighbor;
end j;
for m = 1 to M do:
  for j = 2 to Nⱼ − 1 do:
    receive R(iₛ − 1,j) from west neighbor;
    for i = iₛ to iₑ do:
      calculate ρ(i,j), R(i,j);
    end i;
    send R(iₑ,j) to east neighbor;
  end j;
  for j = Nⱼ − 1 to 2 step −1 do:
    receive δ(iₑ + 1,j) from east neighbor;
      for i = iₑ to is step −1 do:
        calculate δ(i,j);
        update variable
    end i;
    send δ(iₛ,j) to west neighbor;
  end j;
end m
```

The problem is that this parallelization technique requires a lot of (fine-grained) communication and there are idle times at the beginning and end of each iteration; this reduces efficiency. Also, the approach is limited to structured grids. Bastian [4] obtained good efficiency on transputer-based machines, which have a favorable ratio of communication to computation speed. With a less favorable ratio, the method would be less efficient.

The transputer was a revolutionary processor architecture of the 1980s that consisted of integrated memory and serial communication links, intended for parallel computing. It was designed and produced by an English company, Inmos, located in Bristol, the United Kingdom [5].

Considerably, the iterative methods are efficient in another way that they are for non-linear problems, but they are just as valuable for sparse linear systems. In this method, one guesses a solution, and uses the equation to systematically improve it. If each iteration is cheap and the number of iterations is small, an iterative solver may cost less than a direct method. In CFD problems this is usually the case. The conjugate gradient is an example of an iterative solver, it can be parallelized in a straightforward manner. The algorithm involves some global communication (gathering of partial scalar products and broadcasting of the final value) but the performance is nearly identical to that on a single processor. However, to be really efficient, the conjugate gradient method needs a good pre-conditioner. Because the best pre-conditioners are of the ILU type, the problems described above come into play again.

This development shows that most parallel computing environments require redesign of algorithms. Methods that are excellent on serial machines may be almost impossible to use on parallel machines. Also, new standards are now used to assess the effectiveness of a method. This means that good parallelization of implicit methods requires the solution algorithm to be modified. Performance in terms of the number of numerical operations may be poorer than on a serial computer, but if the load carried by the processors is equalized and the communication overhead and computing time are properly matched, the modified method may be more efficient overall, as is done in most commercially available codes.

6.1 Domain Decomposition in Space

Parallelization of implicit methods is usually based on data parallelism or domain decomposition, which can be performed in both space and time. In spatial domain decomposition, the solution domain is divided into a certain number of sub-domains; this is similar to block-structuring of grids. In block-structuring, the process is governed by the geometry of the solution domain, whereas in domain decomposition, the objective is to maximize efficiency by giving each processor the same amount of work to do. Each sub-domain is assigned to one processor, but more than one grid block may be handled by one processor. If so, we may consider all of them to be **one logical sub-domain**. As noted, one has to modify the iteration procedure for parallel machines. The usual approach is to split global coefficient matrix A into a system of diagonal blocks, A_{ii}, which contain the elements connecting the nodes that belong to the ith sub-domain, and off-diagonal blocks or coupling matrices A_{ij} ($i \neq j$), which represent the interaction of blocks i and j. For example, if a square 2D solution domain is split into four sub-domains and the cells are numbered so that the members of each sub-domain have consecutive indices, the matrix has the structure shown in Figure 3.7; five-point stencil discretization is used in this illustration. The method described below is applicable to schemes using larger computational stencils; in this case, the coupling matrices are larger.

For efficiency, the iterative solver for the inner iterations should have as little data dependency (data provided by the neighbors) as possible; data dependency may result in long communication and/or idle times. Therefore, the global iteration matrix is selected so that the blocks are de-coupled, i.e., $M_{ij} = 0$ for $i \neq j$. The iteration scheme for sub-domain i is then:

$$M_{ii}\phi_i^m = Q_i^{m-1} - (A_{ii} - M_{ii})\phi_i^{m-1} - \sum_j A_{ij}\phi_j^{m-1}; (j \neq i) \qquad (3.6)$$

M is the iteration matrix and should be diagonal, tridiagonal or block diagonal. The LU solver is easily adapted to this method. Each diagonal

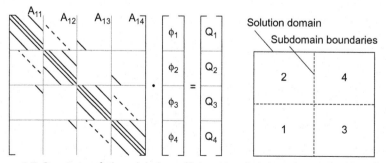

Figure 3.7 Structure of the global coefficient matrix when a square 2D solution domain is divided into four sub-domains.

block matrix M_{ii} is decomposed into L and U matrices in the normal way; the global iteration matrix $M = LU$ is not the one found in the single processor case. After one iteration is performed on each sub-domain, one has to exchange the updated values of the unknown ϕ^m so that the residual ρ^m can be calculated at nodes near sub-domain boundaries.

When the solver is parallelized in this way, the performance deteriorates as the number of processors becomes large; the number of iterations may double when the number of processors is increased from one to 100. However, if the inner iterations do not have to be converged tightly, parallel LU can be efficient because the algorithm then tends to reduce error rapidly in the first few iterations. Especially if the multi-grid method is used to speed up the outer iterations, the total efficiency is as high as 90% (see Schreck and Perić [6] and Lilek et al. [7]).

Conjugate gradient based methods can also be parallelized using the above approach. The concept is based on the fact that it is possible to minimize a function with respect to several directions simultaneously while searching in one direction at a time. This is made possible by a clever choice of directions. Interested readers may search research reference like [8] for conjugate gradient method. A pseudo-code was shown with the pre-conditioned CG solver [3]. Seidel et al. [9] found that the best performance is achieved by performing two preconditioner sweeps per CG iteration on either single or multiprocessors. Results of solving a Poisson equation with Neumann boundary conditions, which simulate the pressure-correction equation in CFD applications, are shown in Figure 3.8. With one pre-conditioner sweep per CG iteration, the number of iterations required for

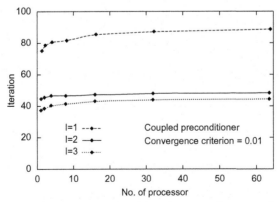

Figure 3.8 Number of iterations in the ICCG solver as a function of the number of processors.

convergence increases with the number of processors. However, with two or more preconditioner sweeps per CG iteration, the number of iterations remains nearly constant. However, in different applications one may obtain different behavior. The code is mentioned below:

- Initialize by setting: $k = 0$, $\phi_0 = \phi_{in}$, $\rho^o = Q - A\phi_{in}$, $p^0 = 0$, $s_o = 10^{30}$
- Advance the counter: $k = k + 1$
- On each sub-domain, solve the system: $Mz^k = \rho^{k-1}$
  - Locally communicate: exchange z^k along interfaces
- Calculate: $s^k = \rho^{k-1} \cdot z^k$
  - GC: gather and scatter s^k
  - $\beta^k = s^k / s^{k-1}$
  - $p^k = z^k + \beta^k p^{k-1}$
  - Locally communicate: exchange p^k along interfaces
  - $\alpha^k = s^k / (p^k \cdot Ap^k)$
  - Global communication: gather and scatter αk
  - $\phi^k = \phi^{k-1} + \alpha^k p^k$
  - $\rho^k = \rho^{k-1} - Ap^k$
- Repeat until convergence.

Here p^k is the search direction, ρ^k is the residual at the kth iteration, z^k is the auxillary vector, α^k and β^k are the parameters in constructing new solution, residual and search direction. To update the right-hand side of Eqn (3.6), data from neighbor blocks are necessary. In the example, processor 1 needs data from processors 2 and 3. On parallel computers with shared memory, these data are directly accessible by the processor. When computers with distributed memory are used, communication between processors is necessary. Each processor then needs to store data from one or more layers of cells on the other side of the interface. It is important to distinguish local (LC) and global (GC) communication.

Local communication takes place between processors operating on neighboring blocks. It can take place simultaneously between pairs of processors; an example is the communication within inner iterations in the problem considered above. Global communication means gathering some information from all blocks in a master processor and broadcasting some information back to the other processors. An example is computation of the norm of the residual by gathering the residuals from the processors and broadcasting the result of the convergence check. If the number of nodes allocated to each processor (i.e., the load per processor) remains the same as the grid is refined (which means that more processors are used), the ratio of LC time to computing time will remain the same.

We say that LC is fully scalable. However, the GC time increases when the number of processors increases, independent of the load per processor. The GC time will eventually become larger than the computing time as the number of processors is increased. Therefore, GC is the limiting factor in massive parallelism. Methods of measuring efficiency are discussed below.

6.2 Domain Decomposition in Time

Implicit methods are usually used to solve steady flow problems. In Fluent, dual-time stepping is used with an implicit scheme so that one-time stepping caters to the iterative procedure whereas the other one solves for the Courant number. Although it was once thought that these methods were not well-suited to parallel computing, they can be effectively parallelized by using domain decomposition in time as well as space. This means that several processors simultaneously perform work on the same sub-domain for different time steps. This technique was first proposed by Hackbusch [10]. Because none of the equations needs to be solved accurately within an outer iteration, one can also treat the old variables in the discretized equation as unknowns. For a two-time-level scheme the equations for the solution at time step n can be written:

$$A^n \phi^n + B^n \phi^{n-1} = Q^n \tag{3.7}$$

Because we are considering implicit schemes, the matrix and source vector may depend on the new solution, which is why they carry index n. The simplest iterative scheme for solving simultaneously for several time steps is to decouple the equations for each time step and use old values of the variables when necessary. This allows one to start the calculation for the next time step as soon as the first estimate for the solution at the current time step is available, i.e., after one outer iteration is performed. The extra source term containing the information from the previous time step(s) is updated after each outer iteration rather than being held constant as in serial processing.

When processor k, working on time level t^n, is performing its mth outer iteration, processor $k-1$, working on time level t_{n-1}, is performing its $(m+1)$th outer iteration. The equation system to be solved by processor k in the mth outer iteration is then:

$$(A^n \phi^n)_k^m + (B^n \phi^{n-1})_{k-1}^m = (Q^n)_k^{m-1} \tag{3.8}$$

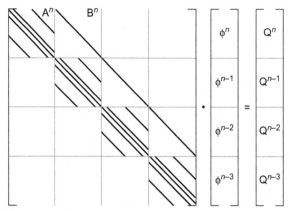

Figure 3.9 Structure of the global coefficient matrix when four time steps are calculated in parallel.

The processors need to exchange data only once per outer iteration; i.e., the linear equation solver is not affected. Of course, much more data are transferred each time than in the method based on domain decomposition in space. If the number of time steps treated in parallel is not larger than the number of outer iterations per time step, using the lagged old values does not cause a significant increase in computational effort per time step.

Figure 3.9 shows the structure of the matrix for a two-time-level scheme with simultaneous solution for four time steps. Time-parallel solution methods for CFD problems were used by Burmeister and Horton [11] and Seidl et al. [9], among others. The method can also be applied to multilevel schemes; in that case, the processors have to send and receive data from more than one time level.

6.3 Efficiency of Parallel Computing

The performance of parallel programs is usually measured by the speedup factor and efficiency, defined by:

$$S_n = \frac{T_s}{T_n} \; ; \; E_{tot} = \frac{T_s}{n T_n} \tag{3.9}$$

Here, T_s is the execution time for the best serial algorithm on a single processor and T_n is the execution time for the parallelized algorithm using n processors. In general, $T_s \neq T_1$, because the best serial algorithm may be different from the best parallel algorithm; one should not base efficiency on the performance of the parallel algorithm executed on a single processor.

Speedup is usually less than n (the ideal value), so efficiency is usually less than 1 (or 100%). However, when solving coupled nonlinear equations, it may turn out that a solution on two or four processors is more efficient than on one processor, so in principle, efficiencies higher than 100% are possible (the increase is often the result of the better use of cash memory when a smaller problem is solved by one processor).

Although not necessary, the processors are usually synchronized at the start of each iteration. Because the duration of one iteration is dictated by the processor with the largest number of mesh cells, other processors experience some idle time. Delays may also result from different boundary conditions in different sub-domains, different numbers of neighbors, or more complicated communication. The computing time, T_s, may be expressed as:

$$T_s = N\tau i_S \tag{3.10}$$

where N is the total number of cells, τ is the time per FLOP, and i_s is the number of FLOPs per cell required to reach convergence. For a parallel algorithm executed on n processors, the total execution time consists of computing and communication time:

$$T_n = T_n^{calc} + T_n^{com} = N_n \tau i_n + T_n^{com} \tag{3.11}$$

where N is the number of cells in the largest sub-domain and T_n^{com} is the total communication time during which calculation cannot take place. Inserting these expressions into the definition of the total efficiency yields:

$$\eta_n^{tot} = \frac{T_s}{nT_n} = \frac{N\tau i_s}{n(N_n\tau i_n + T_n^{com})} = \left(\frac{i_s}{i_n}\right)\left(\frac{N}{nN_n}\right)\left(\frac{1}{1 + \frac{T_n^{com}}{T_n^{calc}}}\right)$$

$$= \eta_{num}\eta_{LB}\eta_{par} \tag{3.12}$$

- η_{num} = The numerical efficiency that caters for effect of the change in the number of operations per grid node required to reach convergence owing to modification of the algorithm to allow parallelization

- η_{LB} = The load balancing efficiency sees for the effect of some processors being idle due to uneven load.
- η_{par} = The parallel efficiency that accounts for the time spent on communication during which computation cannot take place
When parallelization is performed in both time and space, overall efficiency is equal to the product of time and space efficiencies.

Total efficiency is easily determined by measuring the computing time necessary to obtain the converged solution. **Parallel efficiency** cannot be measured directly because the number of inner iterations is not the same for all outer iterations (unless it is fixed by the user). However, if we execute a certain number of outer iterations with a fixed number of inner iterations per outer iteration on one processor, numerical efficiency is unity and total efficiency is then the product of the parallel and load balancing efficiencies. If the load balancing efficiency is reduced to unity by making all sub-domains equal, we obtain parallel efficiency. Some computers have tools that allow operation counts to be performed; then, numerical efficiency can be measured directly.

For both space and time domain decomposition, all three efficiencies are usually reduced as the number of processors is increased for a given grid. This decrease is both nonlinear and problem-dependent. Parallel efficiency is especially affected because the time for LC is almost constant, the time for GC increases, and the computing time per processor decreases as a result of reduced sub-domain size. For time parallelization, the time for GC increases whereas the LC and computing times remain the same when more time steps are calculated in parallel for the same problem size. However, the numerical efficiency will decrease disproportionately if the number of processors is increased beyond a certain limit (which depends on the number of outer iterations per time step).

Optimization of the load balancing is generally difficult, especially if the grid is unstructured and local refinement is employed. There are algorithms for optimization but they may take more time than flow computation.

Parallel efficiency depends on three main parameters:
- Setup time for data transfer (latency time)
- Data transfer rate (usually expressed in megabytes per second)
- Computing time per FLOP (FLOP/s).

For a given algorithm and communication pattern, one can create a model equation to express parallel efficiency as a function of these parameters and the domain topology. Published results [3] showed that parallel efficiency can be predicted to a good extent. One can also model numerical efficiency as a function of alternatives in the solution algorithm, the choice of solver, and the coupling of the sub-domains. However, empirical input based on experience with similar flow problems is necessary, because the behavior of the algorithm is problem-dependent. These models are useful if the solution algorithm allows alternative communication patterns; one can choose the most suitable pattern for the computer used. For example, one can exchange data after each inner iteration, after every second inner iteration, or only after each outer iteration. One can employ one, two, or more preconditioner iterations per conjugate gradient iteration; preconditioner iterations may include local communication after each step or only at the end. These options affect both numerical and parallel efficiency; a tradeoff is necessary to find an optimum.

Combined space and time parallelization is more efficient than pure spatial parallelization because for a given problem size, efficiency goes down as the number of processors increases.

Ferziger and Perić [3] showed results for the case in Table 3.1 involving the computation of unsteady 3D flow in a cubic cavity with an oscillating lid at a maximum Reynolds number of 10^4, using a $32 \times 32 \times 32$ cell grid and a time step of $\Delta t = T/200$, where T is the period of the lid oscillation. When 16 processors were used with four time steps calculated in parallel and the space domain was decomposed into four sub-domains, total numerical efficiency was 97%. If all processors are

Table 3.1 Numerical efficiency for various domain decompositions in space and time to calculate cavity flow with an oscillating lid

Decomposition in space and time	Mean number of outer iterations per time step	Mean number of inner iterations per time step	Numerical efficiency (%)
$1 \times 1 \times 1 \times 1$	11.3	359	100
$1 \times 2 \times 2 \times 1$	11.6	417	90
$1 \times 4 \times 4 \times 1$	11.3	635	68
$1 \times 1 \times 1 \times 2$	11.3	348	102
$1 \times 1 \times 1 \times 4$	11.5	333	104
$1 \times 1 \times 1 \times 8$	14.8	332	93
$1 \times 1 \times 1 \times 12$	21.2	341	76
$1 \times 2 \times 2 \times 4$	11.5	373	97

used solely for spatial or temporal decomposition, numerical efficiency drops below 70%.

Communication between processors usually hampers computation. However, if communication and computation take place simultaneously (which is possible on some new parallel computers), many parts of the solution algorithm can be rearranged to take advantage of this. For example, while LC takes place in the solver, one can perform computation in the interior of the sub-domain. With time parallelism, one can assemble the new coefficient and source matrices while LC is taking place. Even the GC in a conjugate gradient solver, which appears to hinder execution, can be overlapped with computation if the algorithm is rearranged. Convergence checking can be skipped in the early stages of computation, or the convergence criterion can be rearranged to monitor the residual level at the previous iteration, and one can base the decision to stop on that value and the rate of reduction.

7. SOLVING PARTIAL DIFFERENTIAL EQUATIONS

7.1 Hyperbolic Equations

The nature of partial differential equations (PDEs) can best be understood through the concept of a characteristic line. It is a line that forms in any flow field; along this line, the highest-order derivatives of independent variables (usually x, y, and t) are indeterminate over a line and across the line may be discontinuous is called characteristic line. The PDEs are classified on the basis of discriminant values or the derivative values obtained across the characteristic lines. If the value of the derivative across the characteristic line is real and unique, the equation is parabolic. If it is real but not unique, the equation is hyperbolic; and if the value of derivative is imaginary, the PDE is elliptic. For details of the discriminant, the reader may refer to the texts such as Anderson [12] and Ferziger and Perić [3]. If the value of the discriminant is real and distinguished, the PDE is said to be hyperbolic. A wave equation is an example; it is widely studied in classical CFD not only because of its simple nature, but because it is the equation upon which the bricks of initial foundations of CFD are laid. Godunov method, Riemann solver, and TVD schemes usually began with the wave equation as an example. Usually PDEs are solved using either explicit or implicit schemes. We will explain implicit schemes here because they are concerned with computational requirements rather than explicit methods, as they can be parallelized more easily. With implicit methods, the equation after discretization takes the form $\mathbf{AX} = \mathbf{b}$, where $\mathbf{A}$ is the coefficient matrix, $\mathbf{X}$ is the variable column vector, and $\mathbf{b}$ is the column vector of constants.

Often, $\mathbf{A}$ is a sparse matrix; thus, an indirect method or iterative method is used. When $\mathbf{A}$ is dense, a direct method such as LU decomposition or Gaussian elimination is used, as discussed.

7.1.1 One-Dimensional Wave Equation

We first demonstrate a parallel solution of a 1D wave equation with the simplest method, i.e., an explicit finite difference scheme. Consider the following 1D second-order wave equation:

$$u_{tt} = c^2 u_{xx}, \quad \forall\, t > 0 \tag{3.13}$$

We perform finite difference operation (applying central differences on both sides):

$$\frac{u_i^{n+1} - 2u_i^n + u_i^{n-1}}{\Delta t^2} = c^2 \frac{u_{i+1}^n - 2u_i^n + u_{i-1}^n}{\Delta x^2} \tag{3.14}$$

where n is the temporal step and i is the spatial step. After rearranging the terms, we obtain a general equation of the form:

$$u_i^{n+1} = f\left(u_i^{n-1} u_i^n, \ldots, u_{i+1}^n u_{i-1}^n\right) \tag{3.15}$$

After decomposing the computational domain into sub-domains, each processor is given a sub-domain to solve the equation. When performing updates for the interior points, each processor can behave like a serial processor. However, when a point on the processor boundary needs updating, a point from a neighboring processor is needed. This requires communication. Most often, this is done by building buffer zones for the physical sub-domains. After updating its physical sub-domain, a processor will request that its neighbors (two in 1D, eight in 2D, and 26 in 3D) send their border mesh point solution values (the number of mesh point solution values sent depends on the numerical scheme used) and in the case of irregular and adaptive grid, the point coordinates. With this information, this processor then builds a so-called virtual sub-domain that contains the original physical sub-domain surrounded by the buffer zones communicated from other processors (Figure 3.10 and Figure 3.11).

Performance analysis: Suppose M is the total number of mesh points uniformly distributed over P processors. Also, let t_{comp} be the time to update a mesh point without communication and let t_{comm} be the time to

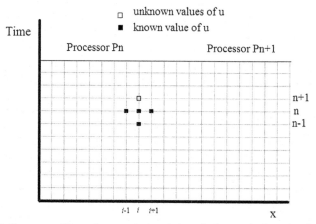

Figure 3.10 Decomposition of computational domain into sub-domains for solving a wave equation.

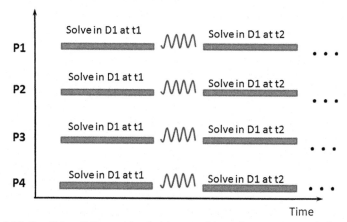

Figure 3.11 Parallel activity graph of decomposition of computational domain into sub-domains for solving a wave equation.

communicate one mesh point to a neighboring processor. The total time for one processor to solve such a problem is:

$$T(1, M) = Mt_{comp} \tag{3.16}$$

With P processors, the time is:

$$T(P, M) = \frac{M}{P} t_{comp} + 2t_{comm} \tag{3.17}$$

Thus, the speedup is:

$$S(P, M) = \frac{T(1, M)}{T(P, M)} = \frac{P}{1 + \frac{2Pt_{comm}}{M \ t_{comp}}} \quad (3.18)$$

and the overhead is:

$$h(P, M) = \frac{2Pt_{comm}}{M \ t_{comp}} \quad (3.19)$$

It is obvious that this algorithm is likely the simplest parallel PDE algorithm. It is common for algorithms for solving PDEs to have such overhead dependencies on granularity and communication-to-computation ratio.

We can easily observe the following:

1. Overhead is proportional to P/M, where M is the number of mesh points on each processor. This means that the more mesh points each processor holds, the smaller the overhead is.
2. Overhead is proportional to t_{comm}/t_{comp}, which is a computer-inherent parameter.

7.1.2 Two-Dimensional Poisson Equation

Consider the following 2D Poisson equation:

$$u_{xx} + u_{yy} = f(x, y) \quad (3.20)$$

Applying central differences on both sides, we get:

$$\frac{u_{i+1,j} - 2u_{i,j} + u_{i-1,j}}{\Delta x^2} + \frac{u_{i,j+1} - 2u_{i,j} + u_{i,j-1}}{\Delta y^2} = f(x, y) \quad (3.21)$$

Assuming $\Delta x = \Delta y = 1$ (without losing generality), we can get the following simplified equation:

$$u_{i+1,j} + u_{i-1,j} + u_{i,j+1} + u_{i,j-1} - 4u_{i,j} = f(x, y) \quad (3.22)$$

We can have the following linear system of algebraic equations:

$$\begin{cases} u_{2,1} + u_{0,1} + u_{1,2} + u_{1,0} - 4u_{1,1} = f(x_1, y_1) \\ u_{3,2} + u_{1,2} + u_{2,3} + u_{2,1} - 4u_{2,2} = f(x_2, y_2) \\ \cdots \end{cases} \quad (3.23)$$

If we define a new index,

$$m = (j-1)X + i \qquad (3.24)$$

to arrange a 2D index (i, j) to a 1D $m = 1, 2, \ldots X, X+1, X+2, \ldots XY$. We can arrange the following 2D discrete Poisson equation:

$$Au = B \qquad (3.25)$$

where A is a sparse matrix defined as:

$$A = \begin{pmatrix} A_1 & I & \cdots & 0 & 0 \\ I & A_2 & I & 0 & 0 \\ \vdots & \vdots & \ddots & \vdots & \vdots \\ 0 & 0 & \cdots & A_{Y-1} & I \\ 0 & 0 & \cdots & I & A_Y \end{pmatrix}_{Y \times Y} \qquad (3.26)$$

This is a $Y \times Y$ matrix with sub-matrix elements

$$A_i = \begin{pmatrix} -4 & 1 & \cdots & 0 & 0 \\ 1 & -4 & 1 & 0 & 0 \\ \vdots & \vdots & \ddots & \vdots & \vdots \\ 0 & 0 & \cdots & -4 & 1 \\ 0 & 0 & \cdots & 1 & -4 \end{pmatrix}_{X \times X} \qquad (3.27)$$

which is an $X \times X$ matrix and there are Y of them, i.e., $i = 1, 2, \ldots$ Y. Then, the final matrix A would be an $(XY) \times (XY)$ matrix. Therefore, the solution of a 2D Poisson equation is essentially the solution of a penta-diagonal (five-diagonal) system.

7.1.3 Three-Dimensional Poisson Equation

Consider the following 3D Poisson equation:

$$u_{xx} + u_{yy} + u_{zz} = f(x, y, z) \qquad (3.28)$$

Applying central differences on both sides, we get:

$$\frac{u_{i+1,j,k} - 2u_{i,j,k} + u_{i-1,j,k}}{\Delta x^2} + \frac{u_{i,j+1,k} - 2u_{i,j,k} + u_{i,j-1,k}}{\Delta y^2}$$
$$+ \frac{u_{i,j,k+1} - 2u_{i,j,k} + u_{i,j,k-1}}{\Delta z^2} = f(x, y, z) \qquad (3.29)$$

Assuming $\Delta x = \Delta y = \Delta z = 1$, we can get the following simplified equation:

$$u_{i+1,j,k} + u_{i-1,j,k} + u_{i,j+1,k} + u_{i,j-1,k} + u_{i,j,k+1} + u_{i,j,k-1} - 6u_{i,j,k} = f(x, y, z)$$
$$\text{(3.30)}$$

We can have the following linear system of algebraic equations:

$$\begin{cases} u_{2,1,1} + u_{0,1,1} + u_{1,2,1} + u_{1,0,1} + u_{1,1,2} + u_{1,1,0} - 6u_{1,1,1} = f(x_1, y_1, z_1) \\ u_{3,2,2} + u_{1,2,2} + u_{2,3,2} + u_{2,1,2} + u_{2,2,3} + u_{2,2,1} - 6u_{2,2,2} = f(x_2, y_2, z_2) \\ \cdots \end{cases}$$
$$\text{(3.31)}$$

If we define a new index m such that

$$m = (k-1)XY + (j-1)X + i \qquad \text{(3.32)}$$

We can linearize a 3D index (i, j, k) by a 1D index (m). Defining $u_m = u_{ijk}$, we obtain the following 3D discrete Poisson equation:

$$Au = B \qquad \text{(3.33)}$$

where A is a sparse matrix defined as:

$$A = \begin{pmatrix} A^1 & I & \cdots & 0 \\ I & A^2 & I & 0 \\ \vdots & \vdots & \ddots & \vdots \\ 0 & 0 & \cdots & A^Z \end{pmatrix}_{Z \times Z} \qquad \text{(3.34)}$$

whose sub-matrix elements are:

$$A_j^k = \begin{pmatrix} a_1^k & I & \cdots & 0 \\ I & a_2^k & \cdots & 0 \\ \vdots & \vdots & \ddots & \vdots \\ 0 & 0 & \cdots & a_Y^k \end{pmatrix}_{Y \times Y} \qquad \text{(3.35)}$$

where $k = 1, 2, \ldots Z$, and where Z is the number of points in Z direction in the above matrix. So, you should have to define a $(Y \times Y)$ matrix for each index in Eqn (3.34). Then, define each index with respect to Y (or j) in the above matrix as in Eqn (3.35); this is a Y × Y matrix. It has sub-matrix elements such as:

$$d_i^j = \begin{pmatrix} d_1^j & I & \cdots & 0 \\ I & d_2^j & \cdots & 0 \\ \vdots & \vdots & \ddots & \vdots \\ 0 & 0 & \cdots & d_X^j \end{pmatrix}_{X \times X} \tag{3.36}$$

We now define the terms of Eqn (3.19) as:

$$d_1^j = \begin{pmatrix} -6 & 1 & \cdots & 0 \\ 1 & -6 & \cdots & 0 \\ \vdots & \vdots & \ddots & \vdots \\ 0 & 0 & \cdots & -6 \end{pmatrix}_{X \times X} \tag{3.37}$$

Thus, A is a seven-diagonal (XY Z) × (XY Z) sparse matrix and the solution of the above equation requires an iterative method.

8. MODERN METHODS OF MESH PARTITIONING

In most flow solvers, mesh partitioning is done in three steps: graph partitioning, nodes (re)ordering, and block generation. An additional step is the merging of output data generated by each subtask (useful in an industrial context for data visualization and post-processing). As shown in Figure 3.12 and Figure 3.13, two strategies can be used: Either the mesh partitioning is done as a preprocessing step and all computing cores execute the same program during computation (SIMD approach) or a master processor is designed to perform the mesh partitioning task before or during calculation (MIMD approach). In the first approach (Figure 3.12), a new problem modified to the desired number of processors is first defined, and parallel calculation is performed using all processors for computation and input–output. However, this partitioning strategy does not allow a dynamic partitioning procedure, which can be interesting for specific applications; and if that happens during calculation, load balancing errors may vary. The second approach (Figure 3.13) is more user-friendly (the mesh partitioning step is hidden to users) and is adapted to dynamic partitioning and (re) meshing (for moving bodies). During computation, the master processor can be dedicated only to input–output or can be used for computation and input–output. However, during the computational steps, the

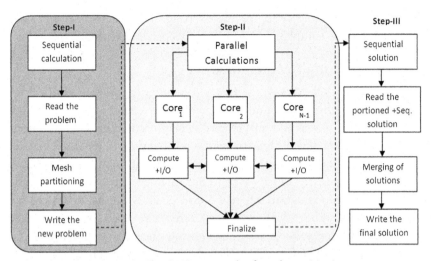

Figure 3.12 The SIMD approach of mesh partitioning.

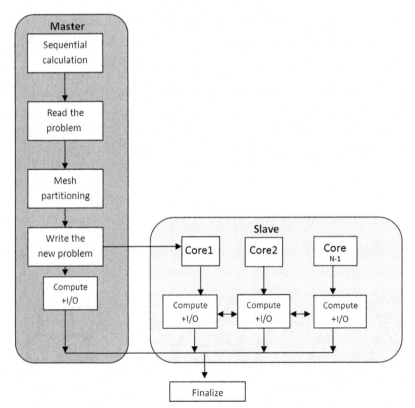

Figure 3.13 The MIMD approach of mesh partitioning.

communication strategy defining the exchange between master and slave processors can be complex and may penalize parallel efficiency.

8.1 Reverse Cuthill–McKee Method

The reverse Cuthill–McKee method is a reordering domain method for unstructured meshes. To consider this method, we have to understand the concept of node graphs and dual graphs. A graph is basically a set of objects connected by some links. These objects in our notation of mesh are vertices whereas the links are edges. Mathematically, a graph is an ordered pair of vertices and edges, i.e., $G = (V,E)$. Thus, a node graph is basically the original mesh representation whereas a dual graph is the sub-mesh created from the node graph. A dual graph has a vertex corresponding to each face of G and an edge joining two neighboring faces for each edge in G. Figure 3.14 represents the node and dual graphs. A sparse matrix is created when the dual graph is partitioned. This partitioning creates sub-domains of vertices and edges that are represented collectively in the form of a matrix. After forming the dual graph, the edges that perform the partition are called edge cuts. It is desirable to have the length of the edge cut as small as possible. Then, the vertices and edges are arranged in a matrix form. The matrix is called an adjacency matrix and depicts which vertices (or nodes) of a graph are adjacent to which other vertices. Thus, the adjacency matrix of a graph with n vertices is the n × n matrix. This is with non–diagonal entry a_{ij} representing the number of edges from vertex i to vertex j, and diagonal entry a_{ii} is either once or twice the number of edges from a vertex to itself. This is shown in Figure 3.15, in which the axes represent the vertices and the black squares depict an edge between vertices. For CFD applications, the order of the elements in the sparse matrix usually affects the performance of numerical algorithms. The objective is to minimize the bandwidth of the sparse matrix (which represents the problem to be solved) by renumbering the nodes of the computational mesh so that they are as close

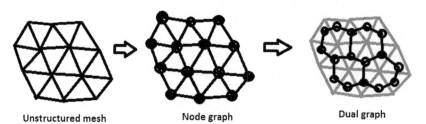

Unstructured mesh   Node graph   Dual graph

Figure 3.14 Node and dual graphs.

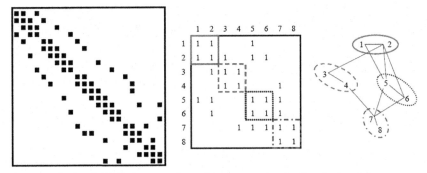

Figure 3.15 Sparse matrix and its connection to the dual graph.

as possible to their neighbors. Thus, an optimal partition would be one that has an equal number of vertices in sub-domains plus the lowest number of edges between sub-domains. This ensures the best use of memory of the computational resources and is obligatory in unstructured CFD solvers.

In ANSYS FLUENT, if we reorder domains, by default it uses the reverse Cuthill–McKee method. It displays the following message when the method is applied:

```
>> Reordering domain using Reverse Cuthill-McKee method:
zones, cells, faces, done.
Bandwidth reduction = 809/21 = 38.52
Done
```

If the bandwidth is printed, the number in the denominator appears, i.e.,

```
Maximum cell distance = 21
```

Thus, the bandwidth is the maximum distance between the neighboring cells (two connected process) in the zone.

8.2 Multilevel or k-Way Partitioning

Multilevel or k-way partitioning is also called METIS partitioning. Multilevel k-way partitioning was devised by Karypis and Kumar [13] and is the default partitioning method in ANSYS FLUENT. It is very fast compared with traditional algorithms for mesh partitioning because it operates with a reduced size graph. Studies have shown that METIS partitioning is better in terms of performance compared with other multilevel algorithms [15]. However, if the performance of METIS is required, the quality of partitioning and the time to partition the mesh must also be considered. Apte et al. [16] mentioned that the best flow technique to evaluate the performance of multilevel algorithms is multiphase flow (two-phase flows). In this scenario, the particles are free to move into the calculation domain. Assuming that the particle density is constant, it then depends only on the volume of the domain. This means that larger volumes would contain more particles. To check the performance of the METIS scheme, we will compare it with the recursive coordinate bisection method (RCB). The RCB algorithm takes into account only information related to the number of cells, whereas the k-way algorithm (METIS) uses more constraints such as particle density. The RCB technique basically partitions the domains so that the sub-domains have equal computational cost. The only problem is with communication performance. It generates long, thin sub-domains that may result in poor performance because the algorithm has to do with communication more than decomposition. This especially affects distributed systems when there is hardware computational overhead. A detailed discussion on this algorithm can be found in Ref. [15]. Gourdain et al. conducted a detailed comparison of the two methods and found METIS to be the fastest. According to Gourdian [2], the RCB technique gives an incorrect solution for load balancing. In contrast, the k-way multilevel algorithm performs better load balancing because of mesh splitting in a reasonable manner. Tests were conducted with a large range of sub-domains, from $64(2^6)$ to $16,384(2^{14})$, to check scalability; METIS took 25 min on partition in a 44 million grid on 4096 sub-domains whereas the RCB algorithm took 270 min on an AMD Opteron-based computer.

However, many studies have proved that METIS gives the fastest results. The exception is for a small mesh or when the number of cores is very large. For the latter case, RCB is comparatively better. Figure 3.16 and Figure 3.17 show the mesh partitioning quality for moderate (up to 10 million cells and 1.9 million nodes) and large (44 million cells and 7.7 million nodes) grids. (Why is the number of nodes are less than the cells?) From a moderate to a very fine mesh, the reduction in the number of nodes and use of the k-way algorithm shows that METIS is considerable. The figures show that the difference becomes larger for a smaller number of nodes when the sub-domains increase. This increases efficiency because of the lower cost of communication and the lower number of nodes.

8.3 Communication between Processes

In CFD, communication is often necessary in multiple data exchange, such as fluxes between block-interfaces or residuals. Thus, it is necessary to

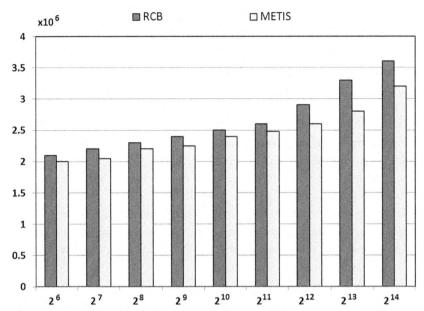

Figure 3.16 Number of nodes after partitioning a combustion chamber simulation [2] using RCB and k-way (METIS) methods with 10 millions cells.

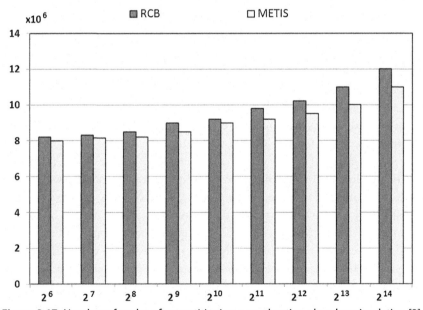

Figure 3.17 Number of nodes after partitioning a combustion chamber simulation [2] using RCB and k-way (METIS) methods with 44 million cells.

provide an application that does this job efficiently. This becomes even more paramount when the number of cores increases to a large extent.

The message passing interface (MPI) is such an application. It is used to communicate data between cores. In MPI there are two types of communication: point-to-point and collective. In point-to-point, the main concern is with data exchange between two particular tasks. An example of this is the MPI send command.

The other type of communication, collective MPI, deals with data exchange among all computing cores involved in a certain task. Whether the communication is point-to-point or collective, the method of doing MPI can involve blocking or non-blocking protocols.

Blocking suspends program execution until the message buffer (a sort of template for message communication) is safe to use. Buffers are slots in computer memory that contain messages for transmission between tasks or to receive replies from a certain task.

Non-blocking routines initiate communication but return even if data have not been successfully sent or received. Thus, the program does not wait to be sure whether the communication buffer is complete.

Gourdian [2] mentioned that although MPI is the most popular commercial CFD code, it is not flexible for new architectures that have more multi-cores and more levels of memory hierarchy.

Open message passing (MP) is different in this regard; it is an open source application. Although it was not designed for distributed architecture, the state-of-the-art API is used on symmetric multiprocessors. It supports multiplatform shared-memory parallel programming in C/C++ and FORTRAN on all architectures.

Open MP has many merits. One is its simplicity because the programmer does not have to wrestle with MP, which is the backbone in the case of writing MPI subroutines. Second, with a little effort, the programmer can choose for which part of the program Open MP is to be used without significant changes in the code. The downside, which I do not think is not a downside because of its design, is that it works on shared memory architecture or symmetric multiprocessing. This is a bottleneck for CFD because CFD codes usually need more than one computing node, but open MP is insufficient to capably parallelize a flow solver whose sole purpose is to solve problems for industrial use. An MP stage is thus a step in which all processors actively participate. Each communication stage is indicated by a list that contains graph edges. These edges are independent. If one edge is taken out, the processors related to the corresponding vertices of this edge will not appear elsewhere. It is also important to structure the MPI programming so that all processes carry an equal amount of messages and the scheduler can assign to them the same number of communication stages.

REFERENCES

[1] Keyes DE, Kaushik DK, Smith BF. Prospects for CFD on Petaflops Systems. Technical Report TR-97-73, Institute for Computer Applications in Science and Engineering 1997.

[2] Gourdian, et al. High performance parallel computing of flows in complex geometries – Part 1: Methods. Computational science and discovery, vol. 2; 2009.

[3] Ferziger Joel H, Perić Milovan. Computational methods for fluid dynamics. 3rd ed. Germany: Springer; 2002.

[4] Bastian P, Horton G. Parallelization of robust multi-grid methods: ILU factorization and frequency decomposition method. In: Hackbusch W, Rannacher R, editors. Notes on numerical fluid mechanics, vol. 30. Braunschweig: Vieweg; 1989. p. 24–36.

[5] http://en.wikipedia.org/wiki/Transputer, accessed in December 2014.

[6] Schreck E, Perić M. Computation of fluid flow with a parallel multi-grid solver. Int J Numer Methods Fluids 1993;16:303–27.

[7] Lilek Ž, Schreck E, Perić M. Parallelization of implicit methods for flow simulation. In: Wagner SG, editor. Notes on numerical fluid mechanics, vol. 50. Braunschweig: Vieweg; 1995. p. 135–46.

[8] Golub GH, van Loan C. Matrix computations. Baltimore: Johns Hopkins University Press; 1990.

[9] Seidl V, Perić M, Schmidt S. Space- and time-parallel navier-stokes solver for 3D block-adaptive cartesian grids. In: Proc. Parallel CFD '95 conference, Pasadena; June 1995.

[10] Hackbusch W. Parabolic multi-grid methods. In: Glowinski R, Lions J-R, editors. Computing methods in applied sciences and engineering. Amsterdam: North Holland; 1984.

[11] Burmeister J, Horton G. Time-parallel solution of the Navier-Stokes equations. In: Proc. 3rd European multigrid conference. Basel: Birkh/iuser Verlag; 1991.

[12] Anderson JD. Computational fluid dynamics-the basics with applications. McGraw Hills; 1996.

[13] Karypis G, Kumar V. Multilevel algorithms for multi-constraint graph partitioning. Tech. Rep. 98–019. USA: University of Minnesota, Department of Computer Science/Army, HPC Research Center; 1998.

[14] Apte SV, Mahesh K, Lundgren TA. Eulerian-Lagrangian model to simulate two-phase particulate flows. Annual research briefs. Stanford (USA): Center for Turbulence Research; 2003.

[15] Hendrickson B, Leland R. A Multilevel algorithm for partitioning graphs. Tech Rep.SAND93-1301. Albuquerque (USA): Sandia National Laboratories; 1993.

[16] Bamford T., August 24, 2007. An implementation of domain decomposition by recursive bisection (M.Sc. thesis). The University of Edinburgh, UK.

CHAPTER 4

High Reynolds Number Flows

1. UNIVERSALITY OF TURBULENCE

Turbulent flow simulations are helpful in CFD and are used to determine the nature of complex turbulence in the case of a high Reynolds number and massively separated flows. Before the seventeenth century, there was no theory regarding the complex fluid flow phenomenon of turbulence. Previous to that, Leonardo De Vinci drew only a conceptual picture of flow behind an obstacle that resembled a blob of hair-like structures. At the end of the seventeenth century Osborne Reynolds set up an experiment and arrived at an important parameter to determine or quantitatively express turbulence. The parameter was devised with his own name, i.e., the Reynolds number, which was found to be a benchmark to classify the nature of flow, whether laminar or turbulent. Kolmogorov, of Russia, conducted parallel work with different experimental studies on turbulence. However, all agreed that turbulence is an uncertain, unsteady, and non-isotropic phenomenon.

After the advent of supercomputers in the mid–1950s, effort was focused on capturing laminar flows numerically. However, in the late 1970s and mid–1980s, work was started on developing turbulence models at a fast rate; advance turbulence schemes were slow paced owing to the unavailability of massively parallel supercomputers. The reason why the need for computational resources increased when turbulence had begun to be solved numerically will be discussed shortly.

Capturing turbulence numerically is indeed a difficult task. Laminar flow equations contain the simple terms of velocity, pressure, and density. In the turbulence world, nothing is simple. The terms of velocity or density are split into two parts: i.e., the mean component and the fluctuating component. It is the fluctuating component in the Navier–Stokes equation that plays the trick. There are five Navier–Stokes equations: there are three velocity component terms and one is continuity and one is the energy equation. Including turbulence means adding more terms to inculcate fluctuating velocity terms. These fluctuating terms are time-averaged, i.e., taking integral over time per unit time. The product of velocities are then manipulated according to averaging rules. This formulation in the Navier–Stokes equations generates the Reynolds-Averaged Navier–Stokes (RANS). The modified form of Navier–Stokes containing the product of fluctuating terms of velocity components is called the Reynolds stresses. Nine stress terms are generated, of which three are symmetrical; thus, six unknown stresses need to be solved. To solve these directly, six equations are required. Therefore, to solve turbulence problems with the RANS method, a lot of data sharing between processor and memory per time iteration is required. Sometimes the memory is not enough to store all of the data at all the grid points, so most software gives memory errors while running in a serial-type computation solver for large meshes. Indeed, the mesh requirement increases because we need to capture turbulent flow.

It has been said that turbulence requires extra equations to solve for turbulent stresses. However, many researchers and scientists have developed a modeling approach to solve turbulence stresses. This approach began with Boussinessq approximation, which related turbulent stress to turbulent kinetic energy. The models were called algebraic or zero equation. Later, scientists such as Wilcox (1988) [1] and Menter (1994) [2] modeled the kinetic energy term with their experiments and observations and developed suitable models to solve RANS. Those were two equation models that solved kinetic energy and dissipation terms. The Launder and Spalding model in 1974 was a k-epsilon model that solved equations for kinetic energy per unit mass and dissipation rate ε. The Wilcox model in 1988 [1] was k-ω and solved for k and specific dissipation. In the era of Baldwin and Barth [3], the thin boundary layer model came into existence specifically for high-speed aerodynamic flows. Spalart Allmaras developed one equation model for modified turbulent viscosity. In this way, some efforts were proven to be useful for generating realistic results with less computer effort; however, services to CFD were still going on to arrive at the real physics of turbulence. The past decade saw advance turbulence modeling in which, with Pentium and Core technology, tera-scale high-fidelity CFD simulations were possible. Thus, the Large Eddy Simulation (LES), and, for simpler cases, the Direct Numerical Simulation (DNS) are now possible to run on medium sized clusters. By medium, we mean a cluster of at least 100 cores.

1.1 What Are LES and DNS?

Most turbulent flows contain eddies. These eddies form, stretch, and then dissipate into static (less energetic) flow. While mixing continues, the phenomenon is followed by the formation of large eddies that continue to deform into smaller ones, and then these deform into much smaller ones until they disspate into heat. This energy transfer mechanism was rhythmically explained by L.F. Richardson [4] as:

Big whorls have little whorls, which feed their velocity; little whorls have lesser whorls, and so unto viscosity.

When large-scale energy eddies break into smaller ones, they have a certain scale of length and time. This scale is based on the time span. However, the energy plot is explained in terms of frequency, which is more sensible and understandable. The plot is called the energy spectrum.

The energy spectrum is the plot of energy levels of eddies that form and dissipate continuously in a turbulent flow regime (Figure 4.1). At each point, this level can be measured experimentally and the decay of turbulent structures (eddies) can be determined. Large eddies (whirls) are energy-containing eddies that dissipate into smaller ones.

To analyze the energy spectrum fully, experiments are the best choice. However, high Reynolds number flow experiments are costly. Therefore, numerical methods are used to examine the energy spectrum. To capture the full spectrum requires capturing the complete energy spectrum. The smallest scales are the most difficult to capture and the mesh size must be on this scale. Scale means the diameter of an eddy in the form of a perfect circle. The smallest is called the Kolmogorov scale. It is of paramount importance to know that, the regions with a certainty of rapidly varying flow structures and that are deemed to contain small eddies, should contain cell size less

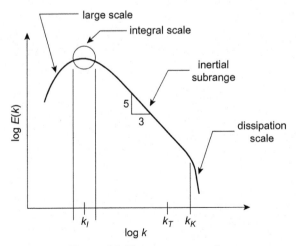

Figure 4.1 The energy cascade.

than or equals to the Kolmogorov scales. Ideally, mesh cells must be of the same small length (the length of the smallest-scale eddies). Practically, this is not possible: If the smallest cells are not of atomic size, but if they continue to spread in a domain of $1 \times 1 \times 1$ m, imagine how many cells would be required. Considering a cell size of 1×10^{-8} m, a 1-m^3 domain will contain $(1 \times 10^8)^3 = 1 \times 10^{24}$ cells. Currently, no supercomputer can solve such a huge mesh size with reasonable accuracy in a reasonable time. If we solve the whole energy cascade, the technique is called DNS.

Usually for the turbulent flows that are statistically steady (that is, there is no change with respect to time in a particular dataset), steady simulations are performed, in which case the energy spectrum is not usually analyzed. For these types of flows, simulations are called RANS. For engineering purposes, these simulations give reasonable results most of the time. However, if the flow is inherently unsteady, the results could be reasonably good but not accurate, so a time-based RANS simulation should be performed. In that case, if the energy spectrum were analyzed, it is likely that it would not cover all of the eddy-time history because of the time-averaging technique of RANS. A much more rigorous unsteady analysis is needed that covers all scales of the spectrum for energy. Let us see if we can find a middle track. Fortunately, there is a way. The key lies in the technique called LES. As in the energy spectrum large eddies dissipate into smaller ones; LES solves only the large ones up to a certain limit determined by the filters. Smaller eddies are modeled using a sub-grid–scale model. With the advent of modern supercomputers, LES is now possible. It is simple to understand that the mesh must contain cells that should be enough to capture the necessary flow characteristics. If the mesh is coarser, the results will be nonphysical. For the energy spectrum, the smallest eddies are in microns (diameter), so the smallest cell must be at least of the same order of magnitude as the smallest eddy. Thus, if the object length is in meters and a supersonic flow needs to be simulated with an obstacle forming certain wake regions as well, the number of cells would be huge, on the order of approximately 10^{30} or more. Thus, for high Reynolds number flows, we can say that simulations of DNS are not currently possible. Large eddy simulations are a good compromise but still need about 10^6 to 10^{15} cells for a simple two-dimensional aerofoil. Before coming down further to unsteady RANS (URANS), another option is available. It is a hybrid technique of LES and URANS. The technique is detached eddy simulation (DES) and is used to solve URANS for near-wall flows and LES for the outer region. The area between LES and URANS is called the gray area.

A switch is used as a constant to toggle between URANS and LES regions. The technique of DES is about 14 years old. Currently, with the availability of HPC, DES is possible for CFD users. It was first proposed in 1997 and first used in 1999. Mostly it is used for high Reynolds number flows and massively separated flows.

1.2 Modeling Turbulence

The viscous regions in high Reynolds number flows are dominated by turbulence. Instability in the shear layer creates turbulent fluctuations of flow-field properties. It has been mentioned that realistic modeling of fluid flow requires the inclusion of turbulence. Usually in turbulence modeling, eddy viscosity is added into the conservation equations. Eddy viscosity can be modeled by certain methods such as algebraic methods, the k-ε model, k-ω, and Spalart Allmaras.

Algebraic models often produce results for limited cases; for example, the Baldwin–Lomax model solves the Navier–Stokes equation in situations that are not appropriate for the Cebeci–Smith model, because the thickness of the boundary layer is not well-defined.

The Johnson–King model [5] has more accurate predictions for shockwave and boundary layer interactions compared with the Baldwin–Lomax and Cebeci–Smith models. The algebraic models keep the boundary layer a tightly coupled entity; in other words, they are boundary-layer models. That is why these models did not become famous for predicting flows involving shear and mixing layers and detached flows. Because of the limitations of algebraic models, scientists use of transport equation models such as k-ε and k-ω. They are both equations models. k is the kinetic energy whereas ε and ω are the dissipation and specific dissipation rate, respectively. These models are known to provide improved results over algebraic models in many cases. They have shown remarkable performance in cases of shockwave boundary layer interaction and separated flows. One limitation to these models is that they require very high mesh near the wall. Near-wall problems often lead to the use of wall functions that are not well-suited to flows involving separation.

The idea of using a single equation for eddy viscosity modeling was conveyed by Nee and Kovasznay in the late 1960s. However, the model was not a local type because it used a characteristic length that scaled to boundary layer thickness. About after 20 years, Baldwin and Barth proposed the Baldwin–Barth model which used a k-model with some assumptions. By modifying the Baldwin-Barth model, Spalart–Allmaras proposed a one-equation model in the early 1990s. It was derived through empirical relations and was effective for predicting aerodynamics flows. The Spalart–Allmaras model employs certain empirical relationships along with Galilean invariance. The model yields fairly rapid convergence to steady state. The wall and free stream boundary conditions are trivial. Furthermore, the low cost and robustness of the eddy-viscosity transport model made this model attractive compared with algebraic models.

2. DIRECT NUMERICAL SIMULATION

Consider a box whose size L must be large enough to carry large eddy motions; grid spacing Δx must be small enough to resolve dissipative scales. In addition, time step Δt, used to advance the solution, is limited by considerations of numerical accuracy.

The DNS mesh requirement is related to the computational cost. This is the time required for a particular grid to be solved on a particular computer, because time is money. In DNS, things are not straightforward. Pope [6] had a detailed analysis on how much the computation cost depends on the Reynolds number of a turbulent flow. The analogy is based on Fourier mode calculation and the wave number, which was defined as K. It was said that the largest eddies, which are energy-containing eddies, have length scales L_{11} (adopting the same nomenclature as the author used). The largest box size L must be equal to $8L_{11}$. The smallest size of grid spacing Δx must be on the order of 1.5 η (where η is the dissipative scale).

Grid size N^3 will depend on integral length scale L_{11}, the turbulent length scale $(k^{3/2}/\varepsilon)$ (where k = kinetic energy and ε = dissipation), Kolmogorov length scale η, and wave number K. The turbulence Reynolds (Re_L) number is given as: $Lk^{0.5}/\nu$ (ν = kinematic viscosity). Thus, grid size N^3 is proportional to:

$$N^3 \approx 4.4 \, Re_L^{9/4} \tag{4.1}$$

Now we come to the temporal (time-based resolution) requirement. The Courant number (Courant Freidrich Lewy number (CFL)) represents the time taken by the flow to travel from one grid point to the next. The time step must be less than grid spacing. A relation is based on turbulent kinetic energy and can be written as:

$$\Delta t = \Delta x/20k^{1/2} \tag{4.2}$$

The total number of time steps must be equal to the total time divided by the time step size, Δt. The total time here is the turbulent time scale, which is given as $\tau = k/\varepsilon$. Thus, the number of time steps, M, can be written as four times the turbulent time scale:

$$M = 4\tau/\Delta t, \qquad (4.3)$$

which is approximately equal to $120 \, \mathrm{Re}_L^{3/4}$.

2.1 Total Floating Point Operations Per Second

It is practical to find the Floating Points Operations per second (FLOPs) by taking the product of both the spatial and temporal resolutions ($N^3 M$).

Assume 1000 floating point operations are required per mode and per time step. The time in days required to perform a simulation on a gigaflop machine would be:

$$T_G = \frac{10^3 N^3 M}{10^9 \times 60 \times 60 \times 24} \qquad (4.4)$$

On the same account, taking the latest supercomputer, Tianhe-2 and its peak performance as 33 petaflops [7], we assume that if 1000 FLOPs were performed on a machine of 1 gigaflops, then on a machine of 33 petaflops, 1×10^9 operations would have to be performed. In that case:

$$T_G = \frac{10^9 N^3 M}{33 \times 10^{15} \times 60 \times 60 \times 24} \qquad (4.5)$$

This is shown for different Reynolds numbers in Table 4.1.

For a 33-petaflop machine, the same table with 10^9 operations per mode per time step is shown in Table 4.2.

Some interesting results have come up here. We can see that if we run DNS simulation with a turbulence Reynolds number of 96,000, it would take 160 years to complete, even on the current supercomputer. It takes about four to five generations to arrive at the outcome of this simulation. Thus, DNS is still seen as a dream.

2.2 Backward-Facing Step Solution

Direct numerical simulation for backward facing, illustrated in Figure 4.2, was performed by Le, Moin, and Kim [8]. A turbulent boundary layer (of thickness 1.2 h and free-stream velocity U_O) enters at the left-hand boundary, separates at step ($x = 0$), and then reattaches downstream

Table 4.1 Computational time cost measure for DNS simulations on a 1-gigaglop machine

Re_L	N	N^3	M	N^3M	CPU time (days)	CPU time break up
94	104	1,124,864	1.20E+03	1.35E+09	0.015623111	22.49728 min
375	214	9,800,344	3.30E+03	3.23E+10	0.374318694	8.983649 h
1500	498	1.24E+08	9.20E+03	1.14E+12	13.151101	13.1511 days
6000	1260	2E+09	2.60E+04	5.20E+13	601.965	20.0655 months
24,000	3360	3.79E+10	7.40E+04	2.81E+15	32,488.96	90.24711 years
96,000	9218	7.83E+11	2.10E+05	1.64E+17	1,903,775.194	5288.264 years

Table 4.2 Computational time cost measure for DNS simulations on a 33-petaflop machine

Re$_L$	N	N^3	M	N^{3}M	CPU time (33-petaflop machine)	CPU time breakup
94	104	1,124,864	1.20E+03	1.35E+09	0.000473428	0.681736 min
375	214	9,800,344	3.30E+03	3.23E+10	0.011342991	0.272232 h
1500	498	1.24E+08	9.20E+03	1.14E+12	0.398518212	13.1511 days
6000	1260	2E+09	2.60E+04	5.20E+13	18.24136364	0.608045 months
24,000	3360	3.79E+10	7.40E+04	2.81E+15	984.5139394	2.734761 years
96,000	9218	7.83E+11	2.10E+05	1.64E+17	57,690.15738	160.2504 years

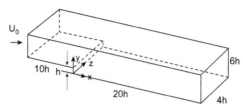

Figure 4.2 Problem schematic of backward-facing step.

(at x $\approx$ 7 h). The flow is statistically stationary and two-dimensional. The Reynolds number considered is Re $= U_O h/v = 5100$. Table 4.3 lists the parameters for the setup of the DNS case.

The Cray C-90 is a vector supercomputer with a processor speed of 4.1 ns (244 MHz). Applying the FLOPs calculation formula, if it were a 16-processors machine, the speed in FLOPs would be:

$$FLOPs = CPUs \times Clock\ speed \times Flops/cycle \qquad (4.6)$$

$$FLOPs = 16 \times 244 \times 10^6 Hz \times Flops/cycle = 4\ GFlops$$

This is purely theoretical formulation to estimate the speed in FLOPs. This is because there are bottlenecks as a result of memory bandwidth, latency of data transfer between processors, etc. We assume that this must be the maximum speed that the Cray C-90 had at that time. Thus, we can say that with 4 gigaflops the simulations took 54 days to complete. It was a great achievement with a supercomputer of that time. The same problem could take a few or 4 days to complete if a modern Cray X1 with 12.8 gigaflops per processor were used, whereas the former had 0.25 gigaflops per processor. In that way, the computational time may be reduced from

Table 4.3 Numerical parameter for DNS simulations of backward-facing step, by Lee et al. [8]

Numerical parameters	Values
Number of nodes in x,N_x	786
Number of nodes in y,N_y	192
Number of nodes in z,N_z	64
Total number of nodes, $N_x N_y N_z$	8.3×10^6
Number of time steps, M	2.1×10^5
Node steps, $N_{xyz}M$	1.8×10^{12}
Computer used	Cray C-90
CPU time (h)	1300

1300 to 25.4 h, or approximately 1 day. You may also evaluate the performance on a Cray XK 6, which is Cray's first graphics processing unit supercomputer [9].

Direct numerical simulation is far more accurate than any numerical method to solve Navier–Stokes equations. It is supportive and assists us with knowledge obtained from experiments of turbulence. It helped us extract Lagrangian statistics that are experimentally impossible to determine [10]. Cases of near-wall flows and homogeneous turbulence can easily be considered in detail from DNS and then through experiments.

3. LARGE EDDY SIMULATION

Large eddy simulation has been the most widely used scale-resolved simulation model over past decades. It is based on the concept of resolving only large scales of turbulence and modeling small scales. The classical motivation for LES is that large scales are problem-dependent and difficult to model, whereas smaller scales become increasingly universal and isotropic and can be modeled more easily.

Large eddy simulation is based on filtering Navier–Stokes equations over a finite spatial region (typically the grid volume) and aimed at resolving only portions of turbulence larger than the filter width. Turbulence structures smaller than the filter width are then modeled—typically by a simple eddy viscosity model.

The filtering operation is defined as:

$$\overline{\Phi} = \int_{-\infty}^{+\infty} \Phi(\vec{x'})G(\vec{x} - \vec{x'})d\vec{x} \tag{4.7}$$

with the term:

$$\int_{-\infty}^{+\infty} G(\vec{x} - \vec{x'})d\vec{x} = 1 \tag{4.8}$$

In this equation, G is the spatial filter. x' is the fluctuating term, and the bar indicates the filtered term. Filtering Navier–Stokes equations results in the following form (considering incompressible flow):

$$\frac{\partial \rho\overline{u}}{\partial t} + \frac{\partial \overline{u}\,\overline{v}}{\partial y} = -\frac{\partial p}{\partial x} + \frac{\partial}{\partial y}\left(\overline{\tau_{ij}} + \tau_{ij}^{LES}\right) \tag{4.9}$$

The equations feature an additional stress term because of the filtering operation, which is τ_{ij}^{LES}. Despite the difference in derivation, the additional sub-grid stress tensor is typically modeled as in RANS using an eddy viscosity model:

$$\tau_{ij}^{LES} = \mu \left(\frac{\partial u}{\partial y} + \frac{\partial v}{\partial x} \right) \tag{4.10}$$

The important practical implication of this modeling approach is that the modeled momentum equations for RANS and LES are identical if an eddy viscosity model is used in both cases. In other words, the derivation of the modeled Navier–Stokes equations is unknown. The only information they obtain from the turbulence model is the size of the eddy viscosity. Depending on that, the equations will operate in RANS or LES mode (or in some intermediate mode). The formal identity of the filtered Navier–Stokes and RANS equations is the basis of hybrid RANS-LES turbulence models, which can obviously be introduced into the same set of momentum equations. Only the model and the numerics have to be switched.

The classical LES models are in the form of the Smagorinsky (1963) [11] model:

$$\mu_t = \rho (C_S \Delta)^2 S \tag{4.11}$$

where Δ is a measure of the grid spacing (filter width) of the numerical mesh, S is the strain rate scalar, given as:

$$S = 0.5 \left(\frac{\partial \overline{u_i}}{\partial y} + \frac{\partial \overline{u_j}}{\partial x} \right) \tag{4.12}$$

and Cs is a constant. This is obviously a simple formulation, indicating that LES models will not provide a highly accurate representation of the smallest scales. From a practical standpoint, detailed modeling might not be required. A more appropriate goal for LES is not to model the impact of unresolved scales on resolved ones, but to model the dissipation of the smallest resolved scales. Large eddy simulation computations are usually performed on numerical grids that are too coarse to resolve the smallest scales. Molecular viscosity is thus not sufficient to provide the correct level of dissipation. In this case, the proper amount of dissipation can be achieved by increasing the viscosity, using eddy viscosity:

$$\varepsilon_{LES} = \frac{\mu_t}{\rho} \left(\frac{\partial \overline{v}}{\partial x} \frac{\partial \overline{u}}{\partial y} \right) \tag{4.13}$$

Eddy viscosity is calibrated to provide the correct amount of dissipation at the LES grid limit. When the LES models are activated, the energy is dissipated and the models provide a sensible spectrum for all resolved scales. In other words, LES does not model the influence of unresolved small-scale turbulence on larger, resolved scales, but rather the dissipation of turbulence into heat (the dissipated energy is typically small relative to the thermal energy of the fluid and does not have to be accounted for, except for high Mach number flows).

3.1 Computational Expense

3.1.1 Free Shear Flows

For free shear flows, LES resolves only the integral length scale (energy-containing eddies). The integral scale varies slowly with the Reynolds number and the computational cost is nearly independent of it.

3.1.2 Wall-Bounded Flows

For the outer layer, $Nx \times Ny \times Nz \sim Re^{0.4}$ and for inner layer $Nx \times Ny \times Nz \sim Re^{1.8}$. The cost would be $Re^{2.4}$. Computationally wall-bounded flows are always expensive [12].

3.1.3 Example 1: Periodic Channel Flow

The dependence on Reynolds number for wall-resolved LES can be estimated for a simple periodic channel flow by taking x direction as stream-wise; y as the wall–normal; and z as the span-wise direction. H is the channel height such that:

$$L_x = 4H, L_y = H = 2h \text{ and } L_z = 1.5H$$

Typical resolution requirements for LES are:

$$\Delta x+ = 40, \Delta z+ = 20 \text{ and } Ny = 60 - 80$$

where $\Delta x+$ is the non-dimensional grid spacing in the stream-wise direction, $\Delta z+$ is in the span-wise direction, and Ny is the number of cells across half the channel height with the definitions:

$$\Delta x+ = \frac{\tau_w \Delta x}{v} \quad (4.14)$$

and

$$\Delta z+ = \frac{\tau_w \Delta z}{v} \quad (4.15)$$

Table 4.4 Number of cells, Nt versus Reynolds number for channel flow [13]

Re	500	1000	10,000	100,000
Nt	5×10^5	2×10^6	2×10^8	2×10^{10}

One can find the number, $Nt = Nx \times Ny \times Nz$, of cells required as a function of the Reynolds number to resolve this limited domain of simple flow (Table 4.4):

$$Nx = 8h/\Delta x = 8Re/\Delta x+, Nz = 8h/\Delta z = 8Re/\Delta z+$$

Reynolds number scaling for channel flows could be reduced by applying wall functions with ever-increasing y+ values for higher Reynolds numbers. However, wall functions are a strong source of modeling uncertainty and can undermine the overall accuracy of simulations. Furthermore, experience with RANS models has shown that the generation of high-quality wall-function grids for complex geometries is challenging. This is even more so for LES applications, in which the user would have to control resolution in all three space dimensions to conform to LES requirements (i.e., $\Delta x+$ and $\Delta z+$, and then depend on y+).

3.1.4 Example 2: LES of Turbulent Jet Flow and Heat Transfer

This example is related to the problem of gas turbine cooling jets. When gas turbine blades are prone to very high temperatures the blade material may be affected. The jets are used within the blade to cool down the temperature. According to the study of [14]; these jets are highly turbulent in nature and it is important to study the turbulent flow phenomenon before its final implementation into the blade. Neumann and Naseem [14] performed a simulation on gas turbine cooling jets with LES. Turbulent impingement with and without swirl on different grids was analyzed and discussed. The jet's Reynolds number was 23,000 and the jet outlet-to-target wall distance was 2. Agreement between experimental data and simulation results was improved by grid refinement in the free shear layer region behind the nozzle tip. The researchers investigated the correlation between the heat transfer mechanism, flow kinematics, and turbulence quantities.

The simulations were performed on two supercomputer clusters available in Stuttgart, Germany. One grid, Grid-I, which had a mesh size of 5 million, was simulated on the Cray Opteron Cluster; then, Grid-II, with 10 million cells, was simulated on the NEC vector platform. Both used the

Table 4.5 Cluster details used for calculations [14]

Cases	CPUs utilized	Cluster used
Coarse grid (5 million)	20	Cray Opteron
Fine grid (10 million)	7	NEC SX-8

solver code FASTEST. The results mentioned that the code FASTEST has an interesting feature with respect to HPC: it has vectorization of subroutines related to the LES sub-grid scale model. The number of processors used for each case is listed in Table 4.5.

The researchers concluded that this investigation was a good comparison of the mean values of velocity and temperature using a coarse grid; they suggested that improved results for correlations of velocity fluctuations could be obtained using Grid-II. Coherent structures appearing as a single jet stream and ring vortex system impingement coincided with the time-dependent flow features that produce high turbulent kinetic energy. This also caused the ring to affect oncoming jet flow.

These simulations were performed on excellent machines. As the name indicates, NEC SX-8 was manufactured by NEC, a Japanese information technology company. SX-8 contains a symmetric multiprocessing system in a compact node module along with a single chip vector processor. It runs at 2 GHz for vector computation and 1 GHz for scalar computations. It can operate at 16 gigaflops and can address up to 128 GB of memory. It has eight CPUs in a single node, and extension may allow the use of 512 nodes. Nevertheless, it seems strange that the author [14] did not use all eight nodes. A multiple of nodes, 2n, is always recommended for better performance. With 512 nodes, the SX-8 system has the model name of 4096M512, in which the prefix indicates the number of CPUs whereas the suffix indicates the number of nodes. It has a peak performance of 65 teraflops. Memory bandwidth is 512 GB/s.

Cray Opteron is a Cray series supercomputer equipped with an AMD Opteron processor. The system that used this AMD microprocessor was Cray XT5, launched on November 6, 2007. It was an updated version of Cray XT4. Cray XT5 has eight sockets supporting dual or quad core AMD Opteron processors.

4. DETACHED EDDY SIMULATION

Detached eddy simulation is a special technique that is useful in cases where the user does not have enough resources to perform heavy simulations with

LES or with DNS. In this case, DES provides a good compromise between RANS and LES. Detached eddy simulation emerged about 18 years ago. Nowadays, with the availability of high-performance computers, DES is practical for CFD users. It was first proposed in 1997 and first used in 1999. Mostly it is used for high Reynolds number flows and massively separated flows. The concept of grid spacing for DES is ambiguous; the performance of DES for a given grid may be less accurate than the performance of DES on a coarser grid. This is because of the crucial requirement of grid spacing for DES simulations, which must be taken into account.

The DES approach proposed by Spalart and co-workers has two basic components: use of LES techniques for free-shear layers or massively separated flows far from solid surfaces. It is difficult to model detached eddies when using RANS; and use of RANS techniques for attached boundary layers.

It is expensive to resolve the eddy structure within these layers using LES; therefore, the effect of these eddies on time-dependent flow should be modeled using RANS to reduce computational cost (Figure 4.3).

The Spalart–Allmaras model functions as a sub-grid model when operating in LES mode and as a wall-layer closure when operating in RANS mode. The form of the governing equations and the method of integration are the same as for unsteady RANS simulations. The turbulence model is modified so that it reverts to a sub-grid model (lower eddy viscosity) in regions where LES is employed and to an RANS model in regions where RANS is employed.

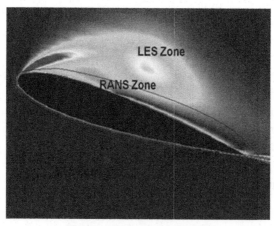

Figure 4.3 The DES concept.

4.1 Computational Requirements

Computational requirements for DES are not stringent as in the cases of LES and DNS. The current author did not find a significant study with regard to the computational requirement for DES alone; however, when solving the problem of DES, many researchers have mentioned the computer resources used for their specific problems. Detached eddy simulation was mostly used for massively separated flows. This typically included flow over a cylinder, aerofoil flow, and backward-facing flow, which were studied and investigated in research such as that of [1,3,15,16].

4.1.1 Flow Around Sphere: Old Study

Dimitris et al. [15] ran a simulation of DES to predict the flow around a sphere. A Mach number of 0.2 was applied and the Reynolds number was 1×10^4 (corresponding to the laminar boundary layer separation region). Along the surface, the mesh spacing was 10^{-4} sphere diameters whereas the domain extent was up to 100 sphere diameters. This made the mesh up to 767,000 vertices. Four multi-grid levels were used and each physical time step was solved with 25 multi-grid cycles with 0.1 and 0.05 time step sizes. Computational cost was 6 min to solve two order of magnitude reductions of 25 multi-grid cycles on 32,800-MHz PIII PCs. Of course, this was achieved on clusters with a technology that is now 12 years old. It was difficult to run DES on a PIII machine at that time. Here, we present a more recent case.

4.1.2 Separated Flow over an Axi-Symmetric Geometry

Sébastein [17] conducted research on the unsteadiness of a separating–reattaching flow field over a geometry that is basically twofold: a backward-facing step (BFS) at the end of which another pipe emanates that contain the nozzle inside it. Thus, the flow is basically a separating flow from the BFS as well as a shear flow interaction owing to the free stream flow interacting with the jet plume. This is an example of massively separated flow in which high-speed flow with a Reynolds number based on the diameter of the body was 1.1×10^6. The free stream Mach number was 0.72. Three different mesh sizes were used; here, the largest one is mentioned from computational point of view: 8.3 million cells. The simulations were run with DES on a single processor of NEC-SX6, which is a prior version of NEC-SX8, as mentioned above. The NEC SX6 is a vector processor whose peak performance is 8 gigaflops. The CPU cost was 1×10^{-6} s/cell per iteration. Time step size was 2 μs. The code was able to perform calculations at 4 teraflops. Total physical time was about 0.2 s. The cost per point and per inner iteration was 1×10^{-6} s. With four inner iterations and a physical time step of 2×10^{-6} s, it needed 4×0.2 s/2×10^{-6} s = 400,000 iterations for a useful calculation to collect statistics.

Total simulated time, including the transient phase to collect statistics, needed close to 600,000 inner iterations. With the finest grid at 8.3 million points, the total CPU cost is 8.3×10^6 points $\times 6 \times 10^5$ inner iterations $\times 1 \times 10^{-6}$ s/point per inner iteration = 1400 CPU hours, i.e., close to 2 months!

The DES simulations took 3 months to complete. The time consumed might have been less than that, but this estimation is based on information provided by Sébastein in [17]. The time might have been less because Sébastein did not specifically mention the number of iterations it took to compute one time step. The author mentioned only the inner loop time iterations, which were four. However, convergence of all equations per time step is important. If all governing equations were converged to second order of accuracy in only four iterations, obviously the total time would have been reduced to less than 1.5 months.

4.1.3 Cavity Flow Study

Gadiparthi [18] conducted a DES study on various cavity geometries. He studied the effect of changing the covered area of the cavity and the trailing edge thickness. Flow field was set at Mach 0.7 and sound pressure levels were obtained from unsteady simulations.

The mesh consisted of six blocks and the total mesh size was 4.1 cells. This was for the finest mesh that Gadiparthi used among the different types of geometries studied. Time step size was 1×10^{-7} and total physical time was 0.05 s. The researcher found that the cavity with the highest trailing edge path had greater Sound Pressure Level (SPL). The simulations were performed at the Wichita High Performance Computing Center. Thirty processors were used for 100 days to perform 500,000 time steps. According to the Web site information for the university [19], the HPC facility contained a dual processor node with Intel Xeon (EM 64 T) 3.6-MHz processors.

Compared with the work of Sébastien, it can be seen that because the processors used in Gadiparthi's simulation were not of high fidelity, the less complex geometry and greater time steps sizes took 100 days (the same time as those in Sébastien's simulation) on the high-performance computer at Wichita University. However, the performance of a vector processor by

NEC-SX6 was far better than the that of the Wichita cluster because in the former a heavy mesh and a small time step size were used.

From this discussion it can be seen that HPC is the correct approach to run very large simulations of DNS and LES or DES. There is no way to compromise if real flow needs to be solved. In principle, DES and LES are suitable for all flows, but they require a substantial amount of preprocessing work to define the corresponding zones and provide suitable grids for all of them. For complex applications, this is not always feasible. It is better to select the safer option over the more convenient one. Direct numerical simulation is still not available for real and practical problems, but it may be that our upcoming generation will see computers than can solve DNS in a couple of days—but surely the destiny is not far enough!

REFERENCES

[1] Wilcox DC. Re-assessment of the scale-determining equation for advanced turbulence models. AIAA J. 1988;26(11):1299–310.
[2] Menter FR. Two-equation eddy-viscosity turbulence models for engineering applications. AIAA J. August 1994;32(8):1598–605.
[3] Baldwin BS, Barth TJ. A One-Equation Turbulence Transport Model for High Reynolds Number Wall-Bounded Flows. 1990. NASA TM 102847.
[4] Richardson LF. Weather Prediction by Numerical Process. Cambridge University Press; 1922. p. 66.
[5] Johnson DA, King LS. A mathematically simple turbulence closure model for attached and separated turbulent boundary layers. AIAA J. 1985;23:1684–92.
[6] Pope SB. Turbulent Flows. USA: Cornell University, Cambridge University Press; 2000.
[7] www.top500.org (accessed 08.05.14.).
[8] Le H, Moin P, Kim J. Direct numerical simulation of turbulent flow over a backward-facing step. J. Fluid Mech. January 1997;330:349–74. http://dx.doi.org/10.1017/S0022112096003941 (About DOI), Published online: 08 September 2000.
[9] http://www.hpcwire.com/2011/05/24/cray_unveils_its_first_gpu_supercomputer/ (accessed 08.05.14.).
[10] Yeung PK, Pope SB. Lagrangian statistics from direct numerical simulations of isotropic turbulence. J. Fluid Mech. 1989;207:531–86.
[11] Smagorinsky J. General Circulation Experiments with the Primitive Equations. Monthly Weather Review March 1963;91(3):99–164.
[12] Lecture Notes on Advanced Turbulence Modeling. Cranfield University; 2009.
[13] Menter FR. Best Practice: Scale-resolving Simulations in ANSYS CFD. Technical Manual Version 1.0. Germany: ANSYS Inc.; April 2012.
[14] Neumann SO, et al. Thermal and flow field analysis of turbulent swirling jet impingement using large eddy simulation. In: Conference on High Performance Computing in Science & Engineering' 08. Germany: Springer; 2008. p. 301–15.
[15] Mavriplis DJ, Pelaez J, Kandil O. Large Eddy and Detached Eddy Simulations Using an Unstructured Multigrid Solver, In: DNS/LES Progress and Challenges-proceedings of the Third AFOSR International Conference on DNS/LES, Report Date: August 2001.

[16] Spalart PR. Young-Person's Guide Simulation Grids. NASA/CR-2001-211032. Seattle, Washington: Boeing Commercial Airplanes; 2001.

[17] Deck S, Thorigny P. Unsteadiness of an axisymmetric separating-reattaching flow: numerical investigation. J. Phys. Fluids 2007;19. 065103.

[18] Gadiparthi, S.K., 2007. Detached Eddy Simulations of Partially Covered and Raised Cavities (M.Sc. thesis). Witchita State University, USA.

[19] http://www.hipecc.wichita.edu/background.html (accessed 31.05.14.).

CHAPTER 5

Cluster Classification

1. CLASSIFICATION OF CLUSTERS

1.1 Classification on the Basis of Architecture

Parallel architecture is based on how processors, memory, and interconnect are laid out and how resources are managed. They can be classified as follows:

1. Symmetric multiprocessing (SMP)
2. Massively parallel processing (MPP)
3. Nonuniform memory access (NUMA)

With a single processor, memory can have difficulty coordinating processor needs, and when processors become multiple the problem multiplies several times. According to the problem of integrating data and the approach adopted by processors, parallel machines can be categorized on the basis of memory accession. The major difference is between the distributed or shared memory. With distributed memory, each processor has its own physical memory (with L1, L2, and L3 cache) and its own address space. Thus, for example, if two processors refer to a variable X, on a shared memory bank they will access the same memory location.

1.1.1 Symmetric Multiprocessors: Uniform Memory Access

A symmetric multiprocessor is a computer with multiple identical processors that share memory and connect via the same bus. All processors share all available global resources (Figure 5.1). Inter-core communication is determined by the bus speed.

In this type of architecture, any memory location is accessible to every processor and there is no latency in accessing the data. The advantage is that parallel programming for this type of architecture becomes simple if any processor accesses any memory location. This is also referred to as uniform memory access. The programming model that works on this principle is called SMP. Symmetric Multiprocessing can be realized in two ways:

1. Modern desktop computers come with a few processors that access shared memory through a single memory bus; for example, Apple manufactures models with 2 hex-core (6-core) processors. Sharing the

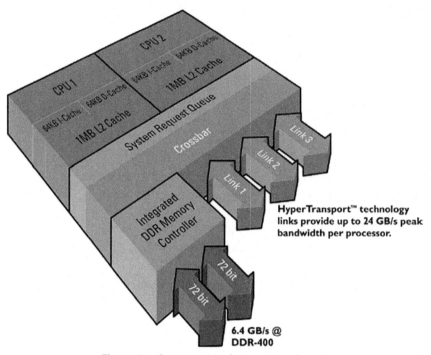

Figure 5.1 Symmetric multiprocessing layout.

memory bus between processors works only for a small number of processors. For multi-core processor, however, the cores typically have a shared cache, such as an L3 or L2 cache.

2. Large numbers of processors use a cross-bar that connects multiple processors to multiple memory banks.

In the early 1960s, Burroughs Corporation introduced a symmetrical multiple-instruction multiple-data multiprocessor with four central processing units (CPUs) and up to sixteen memory modules connected via a cross-bar switch (the first SMP architecture). In the late 1960s, Honeywell delivered the first Multics system, another SMP system of eight CPUs.

While multiprocessing systems were being developed, technology also advanced in the ability to shrink the processors and operate at much higher clock rates. In the 1980s, companies such as Cray Research introduced multiprocessor systems and UNIX-like operating systems that could take advantage of them (CX-OS). With the popularity of uniprocessor personal computer (PCs) systems such as the IBM PC, the late 1980s saw a decline in multiprocessing systems. But currently, more than twenty years later, multiprocessing has returned to these same PC systems through SMP.

1.1.2 Nonuniform Memory Access

Nonuniform memory access is a computer memory design used in multiprocessors, in which memory access time depends on the memory location relative to a processor. With NUMA, a processor can access its own local memory faster than nonlocal memory: that is, memory local to another processor or memory shared between processors. SGI uses NUMA architecture in the Altix 3000 series. In NUMA, the memory access time depends on the memory location relative to a processor. A processor can access its own local memory faster than non-local memory, or memory local to another processor or memory shared between processors (Figure 5.2). One its disadvantage is that the latencies incurred by access to memory require large bandwidth to be supplied to every processor in the network.

In practice, processors are put with local memory along with an exchange network. In this manner, one drawback is that a processor can access its own memory more quickly whereas other processors' memory is slower.

It is challenging for programmer to keep a record of memory location within a processor. Consider two different processors that have a copy of memory location. If one of the processors changes the location in local (cache) memory, this should also be propagated to other processors. This requires synchronization of memory location. Such synchronization is called cache coherence NUMA (NUMA). One problem with ccNUMA is that it is costly with regard to several memory levels. Moreover, ccNUMA may require support from processors and the network, thus making operating system software more complicated. Also, data traveling through the network of processors take up huge bandwidth to make caches coherent.

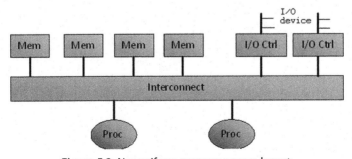

Figure 5.2 Nonuniform memory access layout.

1.1.3 Massively Parallel Processing

An MPP is a single computer with many networked processors. It is a distributed parallel processing (DPP) type of computer that consists of several hundred nodes with a high-speed, specialized interconnection network (Ethernet, Myrinet, or Infiniband). Each node consists of a main memory and one or more processors. Some people distinguish MPP and DPP in that DPP computers are used to solve a single problem. A typical diagram of MPP is shown in Figure 5.3.

What we want is for processors to have their own address space not to see other processors' memory. This approach is called distributed memory (distributed refers to physically as well as logically). If we talk about logically and physically distributed, it means that the processor can exchange information explicitly through the network. Of course, this can be scaled up for a large number of processors. An example is the IBM Blue Gene, which has over 200,000 processors. The downside is that it is the hardest system to program. Hybrids also exist, such as the Columbia computer at the *National Aeronautics and Space Administration (NASA)*, which has twenty nodes connected via a switch; each node is itself NUMA architecture (internal shared memory system) with 512 processors.

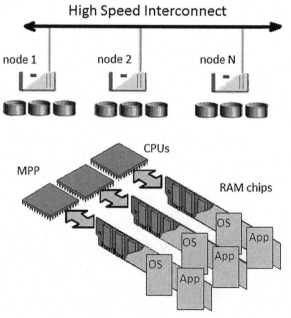

Figure 5.3 Massively parallel processing layout.

In MPP and DPP, several computational entities communicate with the message-passing interface. To use MPP effectively, an information-processing problem must be able to be broken down into pieces all of which can be solved simultaneously. In a scientific environment, certain simulations and mathematical problems can be split apart and each part processed at the same time. In addition, an MPP must be fault tolerant (capacity of bearing a fault) in each individual node or computer.

1.2 Grid Computing

Grid computing typically ties multiple clusters together as an integrated collection of sharable resources, optimized for diverse workloads. Grids often incorporate various computing resources that may be in different geographical locations and scattered across multiple nodes. With grid computing, another layer of complexity is added to sharing resource between clusters, and security and authorization requirements in those clusters are distributed among different owners and groups. Grid computing is also known as cloud computing.

Grid computing can greatly speed up job processing time. HP provides benchmark centers in several locations that enable customers to test new applications and employees to resolve performance problems. However, HP employees do not have full visibility into the benchmark centers, which impedes customer service and makes capital planning and budgeting difficult. To resolve the problem, HP deployed a cloud-like global grid to share its worldwide HPC benchmarking resources with customers and internal staff. Moreover, HP uses the resource management facilities in Platform Computing Load Facility (load-sharing facility (LSF)) software to give HP employees access to information about equipment availability and configurations, irrespective of where the equipment is located.

Airline reservations systems can be considered an example of distributed computing in which travel agents frequently access the server system simultaneously. In this system, because the database has to be accessed from different parts of the world simultaneously, a single server is not sufficient. To handle this system, a remote call procedure is used. A central server calls the query on a remote machine. This remote calling may involve the transfer of data, which must be synchronized. This synchronization plus data handling is the responsibility of the storage area network (SAN). Web servers are an example in which it acts like a single server and handles simultaneous access.

However, because our area of focus is basically on scientific computing, keep in mind that the sort of distributed computing mentioned above is different. Scientific applications usually need parallel programming or hardware because a single piece of hardware is not sufficient to cope with the needs of a particular simulation. Distributed computing, on the other hand, requires mainly data handling. Multiple users access the data and application itself does not require multiple processors to run. Scientific computing needs a fast communication network for parallel needs. Distributed systems, on the other hand, do not need such networking because most of the times the central data are synchronized with other servers. Both in HPC and in business computing, the server has to remain available and operative (24/7/365). However, in distributed computing, a company owner, for example, has full provision to shut it down. To connect to a database server, it does not matter whether the actual server executes the request. This is done through virtualization, in which a virtual server can be spawned on any piece of hardware.

Grid computing has its origin according to the way the computer connected, whether through a local area network or a wide area network. Usually machines are connected in parallel, but they may or may not be owned by different institutions. Currently, grid computing is a means to share resources in the form of scientific data, resource data, and software resources.

Some notable examples are Google, Android, and Amazon. Google has giant network of cloud computing that enables users to type what they want to search, and even help in guessing while the user is typing. This efficient capability of guessing has only been possible due to the presence of strong networking. The computing model used by Google is called MapReduce [1]. It is a combination of data parallelism (mapping) and data accumulation (reducing). In both of these techniques, there is no neighbor-to-neighbor communication that is usually adopted in HPC systems. Google has also developed an application called Software as a Service, which lets the user access application interface built-in software such as Word and Excel through a Web browser (client). In this case, the biggest advantage is that users their own dataset irrespective of the central dataset. Android has examples in the form of GIS and GPS. Amazon also uses the MapReduce technique called Hadoop [2].

Cloud computing also differs in that the data are not maintained by the user. They are centrally administered by the monitor and contain special applications not owned by the user. The user may have to pay even to use or work on such servers. Cloud computing is used mainly for huge amounts of data maintained by large companies such as airlines, Google, and Amazon. Because of its joint integration, cloud computers appear as a single computer to the user. With regard to data handling, cloud computing maintains the record of the user, like Amazon Kindle, which maintains a record as a bookmark of where the reader left off irrespective of whether it was used on phone or PC. These servers are usually used for business companies and information technology (IT) personnel but are not specifically designed for the purpose of HPC. They require no specific hardware and time investment. The universality of Internet, the commodity of hardware, and virtualization are the main factors on which cloud computing relies. Owing to the popularity of cloud computing, scientific clouds with the HPC concept are also in the development stage.

1.2.1 Scenarios of Cloud Resources

The following is a broad classification of usage scenarios for cloud resources.

1. Scaling

    Scaling is a good utility for users to set the cluster according to their needs. This means that the capability and capacity of the computing can be enhanced. A term often used is "platform as a service" (PaaS), which means that the cloud provides a software and development platform. Here, cloud resources are used as a platform that can be expanded based on user demand. If the user is running single jobs and is actively waiting for the output, resources can be added to minimize the output time. This is called capability computing. However, if the user is submitting jobs and the time to complete a job or jobs is not of prime importance, resources can be added only when the queue for the jobs develops. This is called capacity computing. Another term used for computing platforms is "infrastructure as a service," which allows the user to customize an operating system, which is not allowed in the case of PaaS.

2. Multi-tenancy

    As the name multi-tenancy indicates, the resources of a single type of software are allocated to multiple users. Each individual can even customize the software according to his or her needs. The software is not purchased but the user has to pay for usage time.

3. Batch Processing

    The cloud can also be used for batch processing when a user has large amount of data to process. MapReduce computation is one technique of batch processing.

4. Storage

Storage services are provided by most cloud providers, which removes the headache of storing data from the user's end.

5. Synchronization

Currently more than 100 companies provide cloud-based services, well beyond the initial concept of computers for rent. Examples include Google, Amazon, and Netflix. The infrastructure for cloud computing can be interesting from a computer science point of view, because it involves distributed file systems, scheduling, virtualization, and mechanisms for ensuring high reliability.

1.3 Cluster Ingredients

Figure 5.4 shows a jigsaw puzzle. The puzzle is not restricted to the pieces shown; many more pieces can be added to the puzzle. However, the user, who may be an engineer, scientist, or IT specialist, may use this puzzle to obtain an idea about the basic ingredients of a typical cluster system. Details about each component are discussed further.

1.3.1 Preferable Characteristics

1. Applicability

The cluster system must have real-time applicability, which requires the smart use of applications and efficient use of resources.

Figure 5.4 Cluster ingredients jigsaw puzzle.

a. Enhanced Performance

   A cluster should be a system that gives enhanced performance.

b. Enhanced Availability

   A cluster system must be available 24/7/365. Fault tolerance is the terminology used for clusters. In simpler terms, it means that system must be sufficient to bear the faults or continue operating properly in the event of the failure of (or one or more malfunctions within) some of its components. For a system to be fault tolerant, redundancy is kept in these systems. This is not specific to nodes but also to storage, power supplies, and so on.

c. Single System Image

   A cluster should look like a complete system that occupy less space with a comfortable working area. It means that the lesser the space it occupies, the better. Although distributed systems take up a lot of space, shared memory architecture has had a significant role in forming single-system images.

d. Fast Communication

   In cluster networking, intercommunication between nodes remains a bottleneck. This has been controlled to a considerable extent with the introduction of the high-speed Infiniband, but it is still above the range of an economical Beowulf cluster. Normally, Infiniband takes up to 20% of the budget cost of the overall cluster. For academic purposes, therefore, people usually go for common and affordable communication networks such as Ethernet. Myrinet is also a plausible choice and is an intermediate between Ethernet and Infiniband.

e. Load Balancing

   Load balancing has an important role in the performance of a cluster. When multiple users submit a job to a cluster, the load-balancing software intelligently distributes the load to whichever nodes are free. The most popular load balancing is done via a software LSF, which is available in the ANSYS FLUENT but requires software such Portable Batch Scheduler (PBS), which is available for pay, and SGE (Sun Grid Engine) and Ganglia, which are free.

f. System Security

   A key feature of using Linux is that it is highly secure. A system must be secure enough that no intruder can get into it to crash your job. A virus-free environment is also mandatory and in Linux, you do not have to worry about that.

g. Programmability (Optional)

For most clusters, if you are getting a turnkey (ready) solution, you do not need to program anything most times. Commercial codes such as ANSYS FLUENT and ANSYS CFX probably do not need programming unless you use user define functions (in Fluent, for example), or if you are making your own CFD code or some other number-crunching code. Frankly speaking, engineers are mostly afraid of programming and they simply run away by saying "It's the programmer's job!"

2. Use and Application

It is important that you know whether you will achieve some degree of performance if you use a cluster or whether you do not need it. Budget allocation is obviously a main factor to consider after deciding to buy a cluster; however, it is not like a small stationery item that you buy from a shop and start using. Thorough research and investigation are always recommended.

3. Environment

A neat, clean, and temperature controlled environment is necessary for these sophisticated clusters to work best. Cooling load calculations must be properly done to account for heat, ventilation, and air-conditioning issues. Space should be properly allocated so that the racks, shelves, and uninterruptible power supply systems do not become congested.

4. Component Selection for an Integrated System

Usually for defense sector applications, FEA and CFD packages are used; the AMD processor has been used by NASA; however, a number of companies are content with Intel Xeon. AMD processors have a common issue regarding a rise in temperature. Chapter 6 compared their performance.

a. Architecture: 32- or 64-Bit

Do you want to use a 32-bit or 64-bit architecture? When it was first introduced, Itanium was popular, but later, when Intel and AMD started to improve the Pentium series with $\times 86$-64-bit architecture, Itanium's fame faded.

b. Main Memory

The main memory in conventional servers (whether blade- or rack-mounted) is the standard RAM chip called DIMM. Currently, it comes in double data rating (DDR). Thus, one could easily think that there must once have been single data rating or SDR RAM.

However, because technology changes so rapidly, DDR RAM is currently in use. The latest technology uses DDR3, which is more powerful and has fast access. Apparently, physically the structure between DDR, DDR2, and DDR3 differs in the position of the teeth that fix into the RAM slot. Although 8 GB RAM on a single chip configuration is available on the market, it is advantageous to use a 64-bit architecture along with a 64-bit operating system.

c. Network

As discussed earlier, a fast network connection depends on the needs of the user. However, for a heavy application—for example, a 3 to 5 million mesh in ANSYS FLUENT—a 1-GB interconnect is slower and we may need to look for Myrinet or Infiniband. Networks have been discussed in detail in Chapter 7.

d. System Software
   – Operating System: Usually for a cluster environment Linux is commonly used. Although Microsoft is also competitive in the marketing and launched Windows HPC Server 2008, 2012, Linux is still used for cluster-type configurations.
   – Auxiliary Software: Auxiliary software includes software such as MPI libraries (MPICH, OpenMP, MPIHP, etc.), several compilers for C++ or FORTRAN, LINPACK and PLAPACK packages, and job schedulers such MOAB, Ganglia, SGE, and PBS.

e. Storage

Storage is a really big issue if you do not give it importance. If your jobs are not space intensive, you may go for internal storage at the head node. However, this creates problems most times if multiple users are accessing it. Common problems are latency issues and writing a lot of data from multiple users. The remedy for this is to use a separate storage network, often referred to as SAN. The SAN copies the data into a separate hard disk storage system. It also has redundancy, so you do not have to worry if one of the hard drives fails.

f. Stand-alone, Rack-Mounted, or Blade?

It all depends on user needs. Technically, there is not much difference among the three. The same number of nodes with the same architecture (regarding processors, RAM, etc.) in a rack-mounted system would give similar performance as in a blade system. However, regarding appearance, the main advantage is space. If we go from a blade to a stand-alone system, blades occupy the least space because they come in chasses, so you can place two or three chasses in a rack. Blade design is the most costly equipment among the three. The reason is that it has separate power provided by the chassis unit, temperature control devices, and monitoring. As shown in Figures 5.7 and 5.8 a blade system occupies less space than a rack, requires fewer network cables, and consumes less power. A rack-mounted design contains individual nodes mounted on it, so each node occupies considerable space. A node is usually more compact than a normal CPU but a little bit thicker than an Apple laptop. The typical thickness of cluster components is measured in a standard called "U." The nominal thickness of the nodes is 1 U, where 1 U corresponds to 1.75 in. A stand-alone configuration is usually referred to as a workstation, or a group of workstations connected together. Because each workstation is a separate PC, stand-alone configurations occupy lot of space. Figures 5.5–5.8 show three types of cluster fittings. Figure 5.5 shows a desktop cluster. It is offered by Cray, and at one platform you can get the benefit of a mini-supercomputer. Its configuration can easily be seen on the Web. Figure 5.6 shows individual PCs clustered together and adjusted in a rack to occupy minimum space. HP is a leading service provider in the HPC community. Apart from printers, laptops,

Figure 5.5 Cray CX1 desktop cluster.

Figure 5.6 Rack-mounted workstations.

and desktop computers, they offer versatile solutions in the field of HPC. Mainly, they deal with rack-mounted and blade servers. HP is strongly competitive with IBM and Dell in the global market. Figures 5.7 and 5.8 show a chassis of blades with rear and front views. Each chassis consists of several blades and each blade contains a processor, memory, and storage. Several chasses are then mounted together in a cabinet.

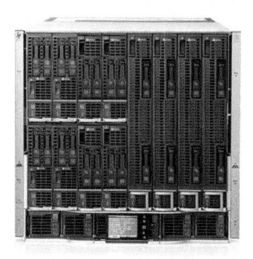

Figure 5.7 Blade server (front view).

Figure 5.8 Blade server (rear view).

2. SOME USEFUL TERMINOLOGY

The reader should become familiar with commonly used terms in HPC. Most terms have been used or described, but they are recalled here for ease of use.

1. Task

    A task is a logically discrete section of computational work. It is typically a program or program-like set of instructions that is executed by a processor. A parallel program consists of multiple tasks running on multiple processors.

2. Pipelining

    Pipelining is breaking a task into steps performed by different processor units, with inputs streaming through, much like an assembly line; a type of parallel computing.

3. Shared Memory

    As the name indicates, it describes a computer architecture in which all processors have direct (usually bus-based) access to common physical memory. This is with regard to hardware. In a programming sense, it describes a model in which parallel tasks all have the same picture of memory and can directly address and access the same logical memory locations regardless of where the physical memory actually exists.

4. Symmetric Multiprocessor

    Symmetric multiprocessing is hardware architecture in which multiple processors share a single address space and access to all resources; shared memory computing.

5. Distributed Memory

    In hardware, this refers to network-based memory access for physical memory that is not common. As a programming model, tasks can only logically see local machine memory and must use communications to access memory on other machines where other tasks are being executed.

6. Communications

    Communication is required for parallel tasks to exchange data frequently. There are several ways this can be accomplished, such as through a shared memory bus or over a network; however, the actual event of data exchange is commonly referred to as communications regardless of the method employed.

7. Synchronization

    Synchronization is the coordination of parallel tasks in real time, often associated with communications. It is often implemented by establishing a synchronization point within an application in which a task may not proceed further until another task(s) reaches the same or logically equivalent point. It usually involves waiting by at least one task, and can therefore cause a parallel application's wall clock execution time to increase.

8. Granularity

In parallel computing, granularity is a qualitative measure of the ratio of computation to communication. Like the CFD mesh, it can also be categorized as

a. Coarse: relatively large amounts of computational work are done between communication events

b. Fine: relatively small amounts of computational work are done between communication events

9. Observed Speedup

Observed speedup of a code that has been parallelized is defined as the ratio between the wall clock time of serial execution and the wall clock time of parallel execution. It is one of the simplest and most widely used indicators of a parallel program's performance. A similar definition is used by ANSYS FLUENT and is discussed in Chapter 6.

10. Parallel Overhead

Parallel overhead is the amount of time required to coordinate parallel tasks, as opposed to doing useful work. Parallel overhead can include factors such as:

a. Task startup time

b. Synchronizations

c. Data communications

d. Software overhead imposed by parallel languages, libraries, operating system, etc.

e. Task termination time

11. Massively Parallel

This refers to the hardware composed of a given parallel system with many processors. The meaning of "many" keeps increasing, but currently, the largest parallel computers can be composed of processors numbering in the hundreds of thousands.

12. Embarrassingly Parallel

Embarrassingly parallel deals with solving many similar but independent tasks simultaneously; there is no need for coordination between tasks.

13. Scalability

This means the ability to demonstrate a proportionate increase in parallel speedup of a parallel system with the addition of more resources. A parallel system may be hardware or software. Factors that contribute to scalability include hardware (particularly memory),

CPU bandwidth and network communication properties, application algorithms, parallel overhead, and related characteristics of your specific application.

REFERENCES

[1] Dean Jeffrey, Ghemawat Sanjay. MapReduce: simplified data processing on large clusters. In: OSDI'04: Sixth Symposium on Operating System Design and Implementation; 2004.
[2] Hadoop Wiki. http://wiki.apache.org/hadoop/FrontPage, last edited 24-10-2014 at 12:27:46 by DevopamMittra, visited on 4th of November 2014.

CHAPTER 6

HPC Benchmarks for CFD

1. HOW BIG SHOULD THE PROBLEM BE?

Computational Fluid Dynamics (CFD) problems are resource-hungry applications. They require a large amount of computer power to be solved. Let us start with the smallest problem. Consider a one-dimensional (1D) partial differential equation. To solve it, you would need at least three points. For its first-order accurate solution, these three points in space and a certain number of points (for example, 300) for time steps are needed. A single core processor such as Pentium IV with 512 MB RAM will easily solve it in a few minutes. Now, add some complexity: make it 2D and go for more grid points. This would need more computational power, such as 1 GB memory. Now, make it 3D and solve the Navier–Stokes equation with 0.5 million grid points. This would add more complexity and you would need at least a dual-core processor and 2 GB RAM. For *Reynolds-averaged Navier–Stokes* (RANS) simulations this is a reasonably good machine. This may not be of paramount importance because for real physics one may need to capture turbulence. For this, it will become mandatory to add turbulence model to the simulation.

For wall-bounded flows, whether internal or external, the mesh is kept fine near the wall to capture near-wall viscous effects, whether turbulent or laminar. From the viewpoint of computational expense, the mesh is usually stretched in terms of geometric progression near the wall, which can make cell size very large at the far field boundaries. Ideally, one should keep all the points equally spaced, but this can make mesh size very fine. Even in the case of mesh stretching when the mesh size has been increased, a dual-core PC would also be useless. The user may require a quad core with 64-bit architecture support so that he or she can use the whole 4 GB of RAM, at least if the operating system is 32-bit. To get maximum power out of your computer, all four cores can be used. In summary, it can be said that high power is required when:

1. There is a 3D problem to solve
2. There is a need to solve turbulence
3. Flows involve flow recirculation, reversed flows, and gradients

Using HPC for Computational Fluid Dynamics
ISBN 978-0-12-801567-4

Table 6.1 Number of cores suitable for a particular mesh size for Fluent simulation

Cluster size (no. of cores)	Fluent case size (no. of cells/mesh size)	No. of simultaneous Fluent simulations
8	Up to 2–3 million	1
16	Up to 2–3 million	2
16	Up to 4–5 million	1
32	Up to 8–10 million	1
32	Up to 4–5 million	2
64	Up to 16–20 million	1
64	Up to 8–10 million	2
64	Up to 4–5 million	4
128	Up to 30–40 million	1
256	Up to 70–100 million	1
256	Up to 30–40 million	2
256	Up to 8–10 million	4
256	Up to 4–5 million	16

4. There is unsteady flow
5. There is a need for direct numerical simulation, large eddy simulations (LES), or detached eddy simulations (DES).

Thus, an obvious question is how big the problem must be to run on a cluster for HPC. Performance benchmarks are done for this purpose. Guidelines based on the number of cores to be used with respect to the problem size are given in Table 6.1. The benchmark was performed on Sun machines [1]. It is common to see performance benchmarks for CFD with mesh size limits crossing 100 million grid points.

2. MAXIMUM CAPACITY OF THE CRITICAL COMPONENTS OF A CLUSTER

2.1 Interconnect

Interconnect sometimes becomes a bottleneck in the performance of a cluster. This becomes tiring especially when one does not understand where the problem is because everything looks fine in terms of functionality. A fast interconnect usually solves the problem. The term "low latency high bandwidth" is used in the networking field; it means that the time for communication between nodes must be as low as possible and the data transfer rate must be high, i.e., large data packets can be transferred in no time. Currently, the largest vendor of Infiniband in the world is the Mellanox.

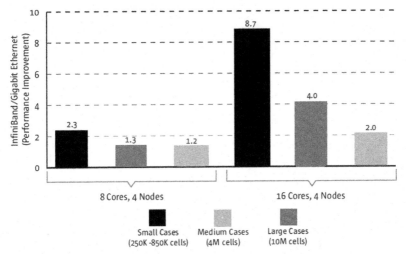

Figure 6.1 Infiniband versus gigabit ethernet.

Figure 6.1 shows a performance histogram for the ratio of Infiniband to gigabit Ethernet. Again, the reference benchmark has been used with respect to the Sun machine [1] with ANSYS® *Fluent*, v. 12 (thanks to Sun, Inc.). The curve has two parts: one is for four nodes each of eight core processors and the other is for four nodes with 16 cores each. The curve is self-explanatory. For a smaller number of cores, a low-size mesh gives subsequent performance but the performance decays and the ratio is increased with the problem size. The ratio of Infiniband to gigabit does not show high values, which indicates that for a low to medium-size mesh (4–10 million), if there is a smaller number of cores, using gigabit Ethernet is sufficient. If the number of cores is increased, Infiniband will have an impact. The first bar has a peak at 8.7, which means that it has an advantage when the number of cores is larger. Consequently, the bars shorten with an increase in problem size because of the large amount of data transfer between nodes. Obviously, the calculations increase as the mesh size increases.

2.2 Memory

Memory requirements for the test cases span from a few hundred megabytes to about 25 GB. As the job is distributed over multiple nodes, the memory requirements per node are reduced correspondingly. As a starting point, 2 GB per core (e.g., 8 GB per dual-processor, dual-core node) is recommended. The total memory requirement for one Fluent 12 simulation on the cluster (distributed across multiple nodes) can scale linearly with

Table 6.2 Estimation of memory requirement for a particular problem size

Fluent case size (no. of cells/mesh size)	RAM requirements
2 Million	4 GB
5 Million	10 GB
50 Million	100 GB

the Fluent model size (measured in number of cells) and can be on the order of the estimates listed in Table 6.2. The author has personally noticed that for steady simulations in almost all versions from 6 to 14, Fluent consumes memory only while reading the case and data files and when it is distributing the mesh over the compute nodes. Figure 6.2 shows the performance of DDR with respect to the software-defined ratio for different meshes on Sun clusters on Fluent 12. The best performance can be obtained with a larger number of cores for a heavy mesh or with an intermediate number of cores on a medium mesh.

2.3 Storage

Adequate storage capacity is also required to run ANSYS Fluent. Data file sizes created by ANSYS Fluent differ with the CFD simulation model size, which is usually measured by the number of cells. With unsteady simulations, the typical data file size increases because of the increasing amount of data to store at each time step. Typical file sizes for a steady case are shown in Table 6.3.

3. COMMERCIAL SOFTWARE BENCHMARKS

3.1 ANSYS Fluent Benchmarks

ANSYS Fluent has several benchmarks available [2]. These benchmarks are versatile in that they contain problems of different scales and have been tested on a number of different platforms. These have been included here so that the reader can have an idea about how the problem depends on the type of hardware used. ANSYS defines benchmarks in terms of the performance rating, speedup, and efficiency. The definition of each term is given below:

1. **Performance Rating:** The performance rating is the basic measure used to report performance results of ANSYS Fluent benchmarks. It is defined as the number of benchmarks that can be run on a given machine (in sequence) in a 24 h period. It is computed by dividing the number of seconds required to run the benchmark by the number of seconds in a day (86,400 s). A higher rating means faster performance.

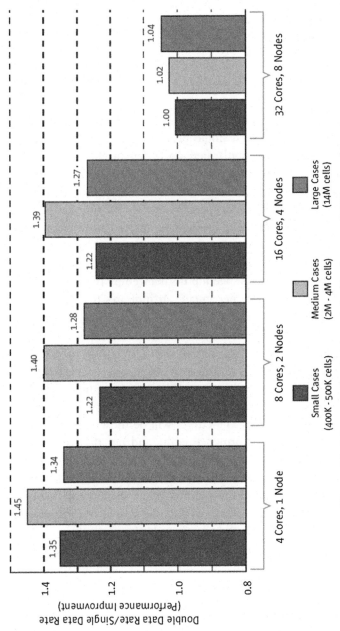

Figure 6.2 Software-defined ratio and double data rate (DDR) comparison.

Table 6.3 Storage needs for a particular problem setup in
ANSYS Fluent

Fluent case size (no. of cells/mesh size)	Space requirements
2 Million	200 MB
5 Million	1 GB
50 Million	5 GB

2. **Speedup:** Speedup is the ratio of wall-clock time required to complete a give calculation using a single processor compared with that of the equivalent calculation performed on a concurrent machine. Its value ranges from 0 to the number of processors used for the parallel run. When speedup is equal to the number of processors used, it is called perfect or linear. Sometimes speedup exceeds the number of processors. This is referred to as super-linear speedup and is often caused by the availability and use of larger amounts of fast memory (e.g., cache or local memory) compared with a single processor run.

3. **Efficiency:** Efficiency is speedup normalized by the number of processors used, presented as a percentage. It indicates the overall use of the central processing units (CPUs) during a parallel calculation. An efficiency of 100% indicates that each CPU is completely occupied by computation during the run period and corresponds to linear speedup. An efficiency of 60% indicates that each CPU is performing useful computation only 60% of the time. The remaining time is spent waiting for other functions, such as parallel communication or work on other processors, to complete. The curves will be shown for the benchmarks with respect to the performance rating and not the standard speedup. The reason is that for speedup performance is measured on the basis of a serial process that is not possible until the benchmark is performed on a single core. Because the technology has made a core to act as a single CPU and in most machines each processor chip contains a minimum of four cores, sometimes it is not possible to measure performance on the basis of a single core, as in most IDM machines. The performance rating thus gives an equivalent measure of speedup with almost the same trend as that depicted in the usual speedup curves.

3.1.1 Flow of Eddy Dissipation

The eddy case consists of the flow modeling of reacting flow with an eddy-dissipation model such as k-ε. Here, ANSYS Fluent has used the k-ε model along with an implicit solver. It has been reported in [2] that this simulation was attempted with 417,000 cells and all-structured mesh. The benchmark results are shown in Figure 6.3. Three machines and their different configurations were used, including the famous hardware of Bull, Fujitsu, and IBM. For the Bull machine, the model was B710 with an Intel E5-2680 processor with a speed of 2.8 GHz with turbo-boost on, and the operating system was Redhat 6 with connectivity of FDR Infiniband by Mellanox. Two models of Fujitsu were tested, each with a difference in CPU speed of 2.7 and 3 GHz, respectively. IBM machines also differ in CPU rating; one is IBM DX 360 M3 with a 2.6 GHz processor and the other is IBM DX360 M4 with a 2.7 GHz processor. All used the same operating system, Redhat Linux 6, and FDR Infiniband of Mellanox for connectivity.

If you look carefully at Figure 6.3, you will see that IBM machine with a 2.6 GHz processor is taking the lead. However, that this machine started at 16 cores at a minimum, which means that the problem is not large enough to require a high number of cores. This is indicated by 256 cores, showing that all of the curves are fading; this is not true with the IBM with a 2.6 GHz processor, which is still a bit straight on 1024 cores compared with others. Table 6.4 mentions other parameters that were described previously. The core solver speedup and core solver efficiency are also

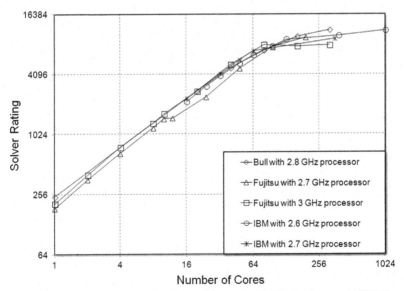

Figure 6.3 Benchmarking curve for the problem of eddy dissipation on ANSYS Fluent 14.5 software.

Table 6.4 Core solver rating, core solver speedup, and efficiency details for the problem of eddy dissipation

Processes	Machines	Core solver rating	Core solver speedup	Core solver efficiency
Bull with 2.8 GHz turbo				
1	1	243.9	1	100%
20	1	2773.7	11.4	57%
40	2	4753.8	19.5	49%
80	4	7125.8	29.2	37%
160	8	9846.2	40.4	25%
320	16	11,443.7	46.9	15%
Fujitsu with 2.7 GHz processor				
1	1	185.4	1	100%
2	1	361.1	1.9	97%
4	1	661.3	3.6	89%
8	1	1185.6	6.4	80%
10	1	1470.6	7.9	79%
12	1	1489.7	8	67%
24	1	2421.9	13.1	54%
48	2	4670.3	25.2	52%
96	4	7697.1	41.5	43%
192	8	9573.4	51.6	27%
Fujitsu with 3 GHz processor				
1	1	206.4	1	100%
2	1	402	1.9	97%
4	1	756.7	3.7	92%
8	1	1318.1	6.4	80%
10	1	1644.1	8	80%
20	1	2769.2	13.4	67%
40	2	5112.4	24.8	62%
80	4	8093.7	39.2	49%
160	8	7819	37.9	24%
320	16	8037.2	38.9	12%
IBM with 2.6 GHz processor				
16	1	2168.1	N/A	N/A
24	2	3083	N/A	N/A
32	2	3945.2	N/A	N/A
48	3	5228.4	N/A	N/A
64	4	6376.4	N/A	N/A
96	6	7783.8	N/A	N/A
128	8	9118.7	N/A	N/A
384	24	9959.7	N/A	N/A
1024	64	11,220.8	N/A	N/A

Table 6.4 Core solver rating, core solver speedup, and efficiency details for the problem of eddy dissipation—cont'd

Processes	Machines	Core solver rating	Core solver speedup	Core solver efficiency
IBM with 2.7 GHz processor				
16	1	2341.5	N/A	N/A
32	2	4148.9	N/A	N/A
48	2	5693.6	N/A	N/A
64	3	6996	N/A	N/A
352	15	9265.4	N/A	N/A

listed. ANSYS Fluent described solver efficiency on one core as 100% because there are no communication bottlenecks and no shared memory headache. Because the solver speedup and efficiency are tabulated with respect to a single core, the columns are shown as not applicable (N/A) for IBM machines, because the minimum number of cores is 16.

3.1.2 Flow Over Airfoil

Figure 6.4 shows benchmarking for the case of flow over an airfoil with about 2 million cells. All of the cells used in the simulation were hexahedral. The turbulence model used was realizable k-ε with a density-based

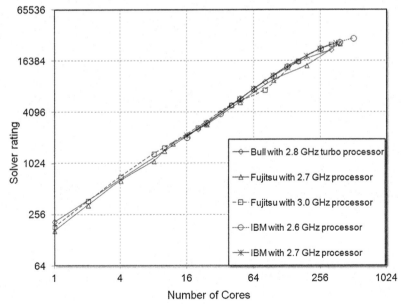

Figure 6.4 Benchmarking curve for the problem of airfoil on ANSYS Fluent 14.5 software.

implicit solver. Based on the solver rating, Fluent scales up very well for a higher number of cores. The problem of 2 million, however, dictates that almost all of the curves start to decelerate after 256 cores. This means that for this problem on all of the machines tested, 256 cores are enough. No significant increase in performance is expected if the cores are extended beyond this number. In comparison, IBM has slightly better performance than Fujitsu and Bull. Table 6.5 shows the results of core solver rating, speedup, and efficiency.

Table 6.5 Core solver rating, core solver speedup, and efficiency details for the problem of flow over aircraft

Processes	Machines	Core solver rating	Core solver speedup	Core solver efficiency
Bull with 2.8 GHz turbo				
1	1	210.3	1	100%
20	1	2583	12.3	61%
40	2	4958.4	23.6	59%
80	4	9340.5	44.4	56%
160	8	15,853.2	75.4	47%
320	16	22,012.7	104.7	33%
Fujitsu with 2.7 GHz processor				
1	1	164.9	1	100%
2	1	329.9	2	100%
4	1	640.4	3.9	97%
8	1	1092.3	6.6	83%
10	1	1442.4	8.7	87%
12	1	1757	10.7	89%
24	1	2921.4	17.7	74%
48	2	5374.8	32.6	68%
96	4	9735.2	59	61%
192	8	14,521	88.1	46%
384	16	25,985	157.6	41%
Fujitsu with 3 GHz processor				
1	1	184.2	1	100%
2	1	368.5	2	100%
4	1	719.1	3.9	98%
8	1	1327.2	7.2	90%
10	1	1575.2	8.6	86%
20	1	2721.3	14.8	74%

Table 6.5 Core solver rating, core solver speedup, and efficiency details for the problem of flow over aircraft—cont'd

Processes	Machines	Core solver rating	Core solver speedup	Core solver efficiency
40	2	4979.8	27	68%
80	4	7500.3	43	50%
160	8	16,225.4	88.1	55%
320	16	25,600	139	43%
IBM with 2.6 GHz processor				
16	1	2076.9	N/A	N/A
24	2	3110.7	N/A	N/A
32	2	3958.8	N/A	N/A
48	3	5877.6	N/A	N/A
64	4	7697.1	N/A	N/A
96	6	11,041.5	N/A	N/A
128	8	14,163.9	N/A	N/A
256	16	22,887.4	N/A	N/A
384	24	27,212.6	N/A	N/A
512	32	30,315.8	N/A	N/A
IBM with 2.7 GHz processor				
16	1	2192.9	N/A	N/A
24	1	3005.2	N/A	N/A
48	2	5798.7	N/A	N/A
64	3	7731.5	N/A	N/A
96	4	11,006.4	N/A	N/A
128	6	13,991.9	N/A	N/A
192	8	18,782.6	N/A	N/A
256	11	22,736.8	N/A	N/A
360	15	26,584.6	N/A	N/A

3.1.3 Flow Over Sedan Car

The problem of a Sedan car was simulated with about 4 million cells. The actual number of cells was 3.6 million. It was a hybrid grid, obviously, owing to the complexity of the geometry especially in the regions where the car wheel is present. k-ε was used as a turbulence model with a pressure–based implicit solver. Figure 6.5 compares benchmarking for this problem; almost all of the machines behave similarly, but the Bull machine is linear until the end and we can expect performance to continue (if not perfectly) if the cores are extended. The behavior is not drifting away from a linear trend, aside from IBM, which shows the curves for the two

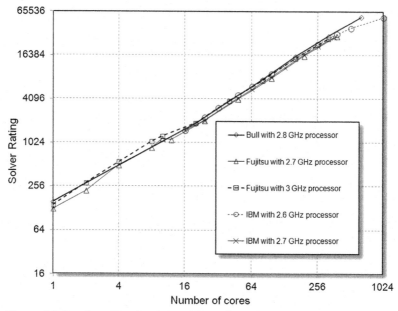

Figure 6.5 Benchmarking for the problem of a sedan car, using ANSYS Fluent.

machines starting to come down at 1024. However, it is expected that after about one x-axis unit all of the curves will start to come down because the problem size is not extravagant. In this problem case, we can say that Bull performed the best. Table 6.6 tabulates Figure 6.5.

3.1.4 Flow Over Truck Body with 14 Million Cells

A truck is an interesting problem from an aerodynamics point of view. Usually, because of their bulky mass, trucks do not attain high velocity on highways. Hence, their drag is reduced by certain geometrical modifications. For example, a fairing is mounted on the roof to reduce drag and thereby increase speed. Also the base drag decreases the speed many fold. Thus, it is the testing through CFD that tells us the contribution of drag on its performance and true simulation can be performed only using techniques like DES. Therefore, HPC is the best possible solution to run these kind of simulations. The benchmark was performed by ANSYS Fluent on a truck body with about a 14-million hybrid type of grid (Figure 6.6). A pressure-based implicit solver was used for simulations. Details regarding the machines used and the efficiency are shown in Table 6.7. Figure 6.6 also shows that the Bull machine performed better than the others. The curve was almost linear until the end.

Table 6.6 Core solver rating, core solver speedup, and efficiency details for the problem of flow over a sedan car

Processes	Machines	Core solver rating	Core solver speedup	Core solver efficiency
Bull with 2.8 GHz turbo				
1	1	157.4	1	100%
20	1	1882.4	12	60%
40	2	3818.8	24.3	61%
80	4	7663	48.7	61%
160	8	15,926.3	101.2	63%
320	16	30,315.8	192.6	60%
640	32	55,741.9	354.1	55%
Fujitsu with 2.7 GHz processor				
1	1	125.9	1	100%
2	1	222	1.8	88%
4	1	502.8	4	100%
8	1	882.3	7	88%
10	1	1175.9	9.3	93%
12	1	1125.4	8.9	74%
24	1	2053.5	16.3	68%
48	2	4085.1	32.4	68%
96	4	7944.8	63.1	66%
192	8	16,149.5	128.3	67%
384	16	29,793.1	236.6	62%
Fujitsu with 3 GHz processor				
1	1	141.2	1	100%
2	1	284	2	101%
4	1	562.1	4	100%
8	1	1066.7	7.6	94%
10	1	1269.2	9	90%
20	1	1916.8	13.6	68%
40	2	3831.5	27.1	68%
80	4	7464.4	52.9	66%
160	8	14,961	106	66%
320	16	27,648	195.8	61%
IBM with 2.6 GHz processor				
16	1	1508.5	N/A	N/A
24	2	2285.7	N/A	N/A
32	2	3091.2	N/A	N/A
48	3	4632.7	N/A	N/A

Continued

Table 6.6 Core solver rating, core solver speedup, and efficiency details for the problem of flow over a sedan car—cont'd

Processes	Machines	Core solver rating	Core solver speedup	Core solver efficiency
64	4	6149.5	N/A	N/A
96	6	9118.7	N/A	N/A
256	16	23,351.4	N/A	N/A
384	24	32,000	N/A	N/A
512	32	38,831.5	N/A	N/A
1024	64	54,857.1	N/A	N/A
IBM with 2.7 GHz processor				
16	1	1519.8	N/A	N/A
24	1	2068.2	N/A	N/A
48	2	4199.3	N/A	N/A
64	3	5610.4	N/A	N/A
96	4	8307.7	N/A	N/A
128	6	11,184.5	N/A	N/A
192	8	16,776.7	N/A	N/A
256	11	21,735.8	N/A	N/A
360	15	29,042	N/A	N/A
16	1	1519.8	N/A	N/A
24	1	2068.2	N/A	N/A

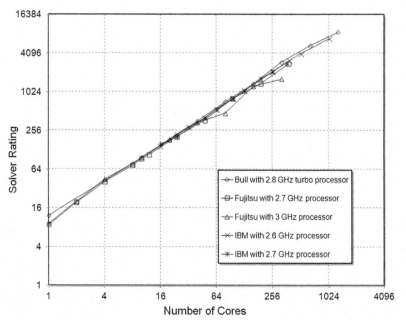

Figure 6.6 Benchmarking of a truck body with 14 million cells using ANSYS Fluent software.

Table 6.7 Core solver rating, core solver speedup, and efficiency details for the problem of flow over a truck body with 14 million cells

Processes	Machines	Core solver rating	Core solver speedup	Core solver efficiency
Bull with 2.8 GHz turbo				
20	1	180.4	14.8	74%
40	2	360.3	29.5	74%
80	4	723.6	59.3	74%
160	8	1373.6	112.6	70%
320	16	2958.9	242.5	76%
640	32	5366.5	439.9	69%
1280	64	8727.3	715.4	56%
Fujitsu with 2.7 GHz processor				
1	1	8.8	1	100%
2	1	19.8	2.2	112%
4	1	41.1	4.7	117%
8	1	74.6	8.5	106%
10	1	93.5	10.6	106%
12	1	108.4	12.3	103%
24	1	206.4	23.5	98%
48	2	371.9	42.3	88%
96	4	811.3	92.2	96%
192	8	1384.6	157.3	82%
384	16	2814.3	319.8	83%
Fujitsu with 3 GHz processor				
1	1	9.3	1	100%
2	1	20.4	2.2	110%
4	1	45	4.8	121%
8	1	79	8.5	106%
10	1	99.3	10.7	107%
20	1	185.9	20	100%
40	2	343	36.9	92%
80	4	483	51.9	65%
160	8	1259.5	135.4	85%
320	16	1624.1	174.6	55%
IBM with 2.6 GHz processor				
16	1	147.2	N/A	N/A
24	2	223.8	N/A	N/A
32	2	292	N/A	N/A
48	3	404.3	N/A	N/A
64	4	548.6	N/A	N/A
96	6	817.4	N/A	N/A

Continued

Table 6.7 Core solver rating, core solver speedup, and efficiency details for the problem of flow over a truck body with 14 million cells—cont'd

Processes	Machines	Core solver rating	Core solver speedup	Core solver efficiency
128	8	1082.7	N/A	N/A
256	16	2112.5	N/A	N/A
384	24	3130.4	N/A	N/A
512	32	4056.3	N/A	N/A
1024	64	6912	N/A	N/A
IBM with 2.7 GHz processor				
16	1	157	N/A	N/A
24	1	203.4	N/A	N/A
48	2	400	N/A	N/A
64	3	539.3	N/A	N/A
96	4	799.3	N/A	N/A
128	6	1057.5	N/A	N/A
192	8	1627.1	N/A	N/A
256	11	2138.6	N/A	N/A
360	15	2851.5	N/A	N/A

3.1.5 Truck with 111 Million Cells

This was the largest benchmark performed by ANSYS Fluent. It consisted of the same problem as discussed before but with 111 million cells. Obviously, with such a huge grid it is difficult to manage the whole grid structure; therefore, mixed-type cells were built. The model was DES turbulence and a pressure base solver was used to solve the governing equations. A perfect linear curve was obtained for Bull machine whereas the worst performance was shown by Fujitsu with a 2.7 GHz processor (see Figure 6.7). Table 6.8 shows the core solver efficiency and speedup.

3.1.6 Performance of Different Problem Sizes with a Single Machine

We have seen that the best machine so far is Bull. We now compare the performance of each problem tested on the Bull machine, shown in Figure 6.8.

Figure 6.8 shows that scalability is good for larger mesh sizes. The smallest mesh of the eddy-dissipation problem, 4000 cells, moves downward after 64 cores. Truck 111 million is still scalable because of the larger mesh size. We can conclude that if one wants to put up a cluster in a laboratory, the first thing is to know the size of the problem size and how large it may be in future. Will you run larger meshes in the future, or

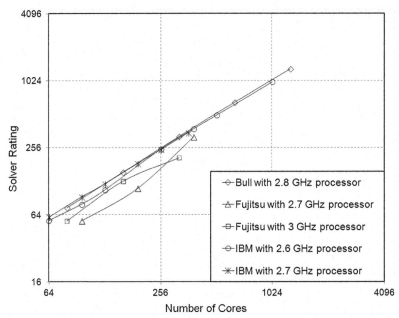

Figure 6.7 Benchmarking of a truck problem with 111 million cells with ANSYS Fluent software.

Table 6.8 Core solver rating, core solver speedup, and efficiency details for the problem of flow over a truck body with 111 million cells

Processes	Machines	Core solver rating	Core solver speedup	Core solver efficiency
Bull with 2.8 GHz turbo				
80	4	74	N/A	N/A
160	8	154.2	N/A	N/A
320	16	322.7	N/A	N/A
640	32	654	N/A	N/A
1280	64	1303.2	N/A	N/A
Fujitsu with 2.7 GHz processor				
96	4	56.3	N/A	N/A
192	8	110	N/A	N/A
384	16	318.5	N/A	N/A
Fujitsu with 3 GHz processor				
80	4	56.3	N/A	N/A
160	8	128	N/A	N/A
320	16	208.2	N/A	N/A

Continued

Table 6.8 Core solver rating, core solver speedup, and efficiency details for the problem of flow over a truck body with 111 million cells—cont'd

Processes	Machines	Core solver rating	Core solver speedup	Core solver efficiency
IBM with 2.6 GHz processor				
64	4	56.7	N/A	N/A
96	6	79.6	N/A	N/A
128	8	107	N/A	N/A
256	16	249.9	N/A	N/A
384	24	378.9	N/A	N/A
512	32	501.7	N/A	N/A
1024	64	998.8	N/A	N/A
IBM with 2.7 GHz processor				
64	3	61.4	N/A	N/A
96	4	93	N/A	N/A
128	6	121.6	N/A	N/A
192	8	185.3	N/A	N/A
256	11	246.7	N/A	N/A
360	15	348.8	N/A	N/A

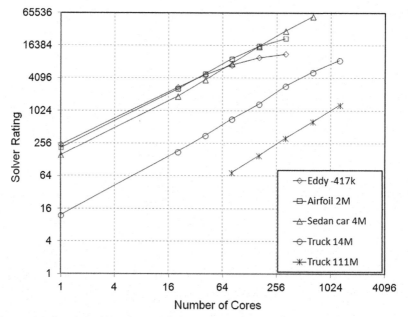

Figure 6.8 Benchmarking for a number of problems on the bull machine with a 2.8 GHz turbo processor.

unsteady simulations? Then, you need to know, if your possible mesh size may be millions of cells, how many cores would be sufficient. Keeping in mind the components of Infiniband, the operating system, and the MPI software constant, these benchmark curves will guide you as to how many cores will be sufficient for your case. The last step is to select the machine. Keeping the budget in mind, you will select a vendor and then compare prices in the market. If a vendor is the best for your problem but is expensive, go to the next best one, and so on. A flowchart will help you select an appropriate machine, as shown in Figure 6.9.

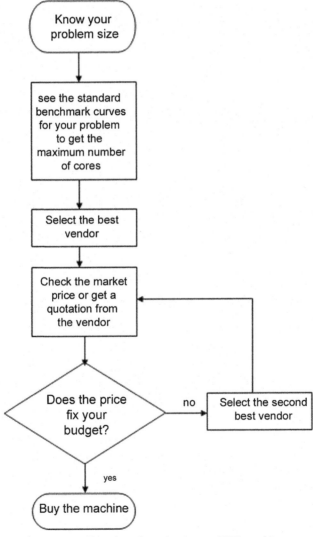

Figure 6.9 Flowchart for selecting an HPC machine.

Alternately, you may obtain quotations from all of the vendors and then compare them and select the best one. You may need to compromise between performance and your budget. Vendors sell their machines mostly on the basis of teraflops per second. This is based on the LINPACK benchmark, but for software such as ANSYS Fluent and CFX or Open-FOAM, you need to look at the number of cores for your problem and choose a machine on this basis. Second, Infiniband is offered by Mellanox, which is the sole vendor in the HPC market, like NVIDIA in GPU technology. Infiniband is expensive equipment, but whatever machine you are going to buy, its price is fixed, so determine whether the vendor is offering it as a package. Otherwise, you will have to make room in the budget for it, as well.

3.2 Benchmarks for CFX

3.2.1 Automotive Pump Simulation

This CFX benchmark was used for an automotive pump problem consisting of 596,252 cells. There were mixed types of elements, including tetrahedrons and prisms. The models used were k-ε and a moving reference frame to inculcate the motion of the rotor and make the stator stationary. A density-based solver was employed in the simulations.

CFX benchmarks differ from ANSYS Fluent benchmarks, in that ANSYS mentions different processor architectures on which the benchmarks are run rather than specifying different vendors. The pump problem was also run with different types of processors, Infiniband architecture, operating systems, etc. The pump problem was mainly run with various Intel processors. The first was with an Intel E5-2670 processor with a 2.6 GHz processor, which had the best performance overall. It had 64 GB RAM per machine and the operating system was Redhat Linux. Next was Intel E5-2680 with a 2.7 GHz processor. This processor was Intel Sandy Bridge, a dual-CPU, 16-core processor with 28 GB RAM and CentOS as the operating system. The last (but not least) was Intel x5650, with a 2.67 GHz processor with 39 GB RAM and SLES as the operating system, as shown in Figure 6.10. The SLES operating system is from SUSE Linux Enterprise Service. This is a much more stable, secure, and user-friendly version of Linux. Figure 6.10 shows the values only for the solver rating; core solver speed and efficiency are listed in Table 6.9.

3.2.2 Le Mans Car Simulation

Millions of dollars are spent improving the design of sports cars. Le Mans is an example. The CFX team performed an analysis fn this car using a mesh

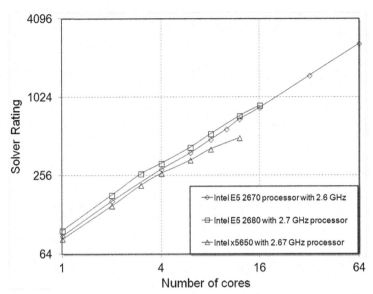

Figure 6.10 Benchmarking for the problem of an automotive pump simulation with ANSYS CFX software.

Table 6.9 Benchmarks performed with CFX software for the pump simulation problem

Processes	Machines	Core solver rating	Core solver speedup	Core solver efficiency
HP SL230sG8 with Intel Sandy Bridge (16-core dual CPU) with 64 GB RAM per machine, RHEL 6.2 using FDR Infiniband without turbo mode				
1	1	88	1	100%
2	1	162	1.84	92%
4	1	285	3.24	81%
6	1	381	4.33	72%
8	1	483	5.49	69%
10	1	580	6.59	66%
12	1	691	7.85	65%
16	1	847	9.63	60%
32	2	1490	16.93	53%
64	4	2618	29.75	46%
Intel Sandy Bridge (16-core dual CPU) with 28 GB RAM				
1	1	96.9	1	100%
2	1	180.8	1.87	93%
3	1	262.6	2.71	90%
4	1	317.6	3.28	82%
6	1	421.5	4.35	73%

Continued

Table 6.9 Benchmarks performed with CFX software for the pump simulation problem—cont'd

Processes	Machines	Core solver rating	Core solver speedup	Core solver efficiency
8	1	533.3	5.51	69%
12	1	732.2	7.56	63%
16	1	881.6	9.1	57%
Intel Gulftown/Westmere (12-core dual CPU) with 39 GB RAM				
1	1	83.4	1	100%
2	1	149.2	1.79	89%
3	1	214.9	2.58	86%
4	1	265	3.18	79%
6	1	334.9	4.02	67%
8	1	407.5	4.89	61%
12	1	496.6	5.95	50%

size of 1,864,025 cells. All of the elements were tetrahedral. The turbulence model was k-ε with a density-based implicit solver. Figure 6.11 shows the benchmark curve and Table 6.10 lists the detailed parameters. Figure 6.11 also shows that Intel E5 2670 depicted a linear trend for even 64 cores, and that this was the best among the three processors tested.

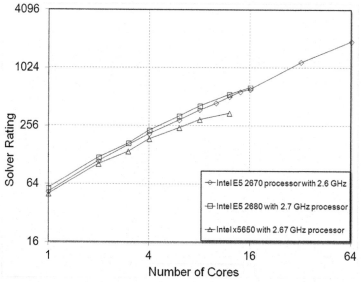

Figure 6.11 Benchmarking for problem of Le Mans car with ANSYS CFX software.

Table 6.10 Core solver rating, core solver speedup, and efficiency details for the problem of flow over a Le Mans car body with ANSYS CFX software

Processes	Machines	Core solver rating	Core solver speedup	Core solver efficiency
HP SL230sG8 with Intel Sandy Bridge (16-core dual CPU) with 64 GB RAM per machine, RHEL 6.2 using FDR Infiniband without turbo mode				
1	1	53	1	100%
2	1	111	2.09	105%
4	1	211	3.98	100%
6	1	294	5.55	92%
8	1	374	7.06	88%
10	1	441	8.32	83%
12	1	514	9.7	81%
14	1	580	10.94	78%
16	1	617	11.64	73%
32	2	1168	22.04	69%
64	4	1920	36.23	57%
Intel Sandy Bridge (16-core dual CPU) with 28 GB RAM				
1	1	57.9	1	100%
2	1	120.8	2.09	104%
3	1	169.7	2.93	98%
4	1	230.4	3.98	100%
6	1	322.4	5.57	93%
8	1	413.4	7.14	89%
12	1	543.4	9.39	78%
16	1	640	11.06	69%
Intel Gulftown/Westmere (12-core dual CPU) with 39 GB RAM				
1	1	50.4	1	100%
2	1	103.1	2.04	102%
3	1	139.1	2.76	92%
4	1	187	3.71	93%
6	1	244.8	4.85	81%
8	1	296.9	5.89	74%
12	1	347	6.88	57%

3.2.3 Airfoil Simulation

Airfoil simulation was conducted with 9,933,000 cells. All of the elements were hexahedral. Shear stress transport was used as a turbulence model whereas coupled implicit simulation was used to solve the flow equations. Figure 6.12 shows the curves and Table 6.11 lists the values of performance parameters.

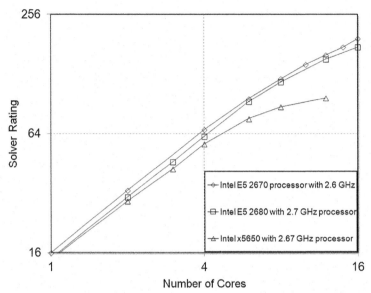

Figure 6.12 Airfoil simulation benchmarking with ANSYS CFX software.

Table 6.11 Core solver rating, core solver speedup, and efficiency details for the problem of flow over an airfoil

Processes	Machines	Core solver rating	Core solver speedup	Core solver efficiency
HP SL230sG8 with Intel Sandy Bridge (16-core dual CPU) with 64 GB RAM per machine, RHEL 6.2 using FDR Infiniband without turbo mode				
1	1	16	1	100%
2	1	33	2.06	103%
4	1	67	4.19	105%
6	1	96	6	100%
8	1	121	7.56	95%
10	1	143	8.94	89%
12	1	159	9.94	83%
14	1	175	10.94	78%
16	1	193	12.06	75%
Intel Sandy Bridge (16-core dual CPU) with 28 GB RAM				
1	1	14.9	1	100%
2	1	30.6	2.06	103%
3	1	46.1	3.09	103%
4	1	62	4.17	104%

Table 6.11 Core solver rating, core solver speedup, and efficiency details for the problem of flow over an airfoil—cont'd

Processes	Machines	Core solver rating	Core solver speedup	Core solver efficiency
6	1	92.9	6.24	104%
8	1	116.6	7.83	98%
12	1	151.8	10.2	85%
16	1	174.9	11.75	73%
Intel Gulftown/Westmere (12-core dual CPU) with 39 GB RAM				
1	1	14.8	1	100%
2	1	29.2	1.97	98%
3	1	42.4	2.86	95%
4	1	56.7	3.82	96%
6	1	76.1	5.13	86%
8	1	87.3	5.88	74%
12	1	96.9	6.53	54%

4. OPENFOAM® BENCHMARKING

Throughout the text we have been discussing commercial codes, mainly ANSYS Fluent. We now focus on open source codes as well. Open source means that you can modify the code according to your need. You may add subroutines, programs, functions, and so on, according to your needs. The most well-known open source code for CFD simulations is OpenFOAM, in which "Open" refers to the open source and "FOAM" stands for Field Operation and Manipulation. The benchmark presented here is for the problem of cavity flow—a famous problem in the CFD community. The case was simulated with the 2.2 version of OpenFOAM.

The following table (Table 6.12) illustrates the conditions under which the problem was simulated.

An Altix system contains processors that are connected by the NUMALINK in a fat-tree topology. The terminology of fat-tree evolves from the fact that branches become thinner as they move from bottom to top; here, network branches become thicker until they reach the master node. Thus, it is like a tree, but inverted.

Like conventional clusters, and in this case as well, each node is fitted into a blade that later fits into an enclosure or chassis; it is also called the individual rack unit (IRU). The IRU is a 10-unit enclosure that contains

Table 6.12 Flow conditions for the cavity flow problem

Reynolds number	1000
Kinematic viscosity	$0.0001 \ \text{m}^2/\text{s}$
Cube dimension	$0.1 \times 0.1 \times 0.1 \ \text{m}$
Lid velocity	1 m/s
deltaT	0.0001 s
Number of time steps	200
Solutions written to disk	8
Solver for pressure equation	Preconditioned conjugate gradient (PGC) with diagonal incomplete Cholesky smoother (DIC)
Decomposition method	Simple

the necessary components to support the blades, such as power supplies, two router boards (one for every five blades), and an L1 controller. Each IRU can support 10 single-width blades or two double-width blades and eight single-width blades. The IRUs are mounted in a 42-U-high rack; thus, each rack supports up to four IRUs. The Altix ICE X blade enclosure features two 4x DDR Infiniband switch blades.

The process was expedited on an SGI machine by Silicon Graphics, Inc®. The model was an Altix Ice X computer with 1404 nodes, each carrying two eight-core Intel Xeon E5-2670 CPUs and 32 GB memory per node. The interconnect was FDR and FDR-10 Infiniband. For this problem, for larger mesh sizes, the case was split into nodes in multiples of 9. You may consider them to be cores. Here, the nodes are grouped in IRUs of 18 nodes each, where each IRU has two switches connecting 9×9 nodes together. It is beneficial to fill these IRUs, with respect to both communication and fragmentation of the job queue.

The simulated cases are shown in Table 6.13. The 27-million mesh size on one node was not run because the RAM was not adequate. On the other hand, smaller mesh sizes are not run on a large number of nodes not because of memory, but because extra time latency in communication causes a drop in parallel performance.

The results of this scaling study are presented as plots indicating speedup and parallel efficiency. All results are based on total analysis time, including all startup overhead.

Speedup and parallel efficiency are calculated with the lowest number of nodes as a reference; i.e., speedup is computed relative to one node for all meshes except the 27-million cell mesh, in which the speedup is relative to two nodes. In Figure 6.13, a trend is evident in that the highest

Table 6.13 Problem scale span and number of cores

Nodes	Mesh size				
	1M	3.4M	8M	15.6M	27M
1N (16 cores)	Yes	Yes	Yes	Yes	No
2N (32 cores)	Yes	Yes	Yes	Yes	Yes
4N (64cores)	Yes	Yes	Yes	Yes	Yes
9N (145 cores)	Yes	Yes	Yes	Yes	Yes
18N (288 cores)	Yes	Yes	Yes	Yes	Yes
27N (432 cores)	Yes	Yes	Yes	Yes	Yes
36N (576 cores)	Yes	Yes	Yes	Yes	Yes
72N (1152 cores)	Yes	Yes	Yes	Yes	Yes
144N (2304 cores)	Yes	Yes	Yes	Yes	Yes
288N (4608 cores)	No	No	Yes	Yes	Yes

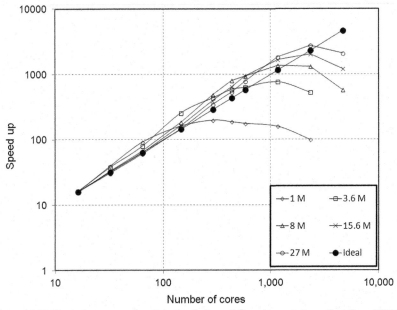

Figure 6.13 Speedup curve for the simulation of flow in a cavity using OpenFOAM.

performance is achieved by 27 million cores. The smallest mesh size case does not show linear behavior because efficiency drops with a higher number of processes as a result of intercommunication, as shown in Figure 6.13. Figure 6.14 plots parallel efficiency. The ideal behavior is obviously one corresponding to 100% efficiency. It is normal to obtain efficient results and then to see efficiency drop after a certain number of

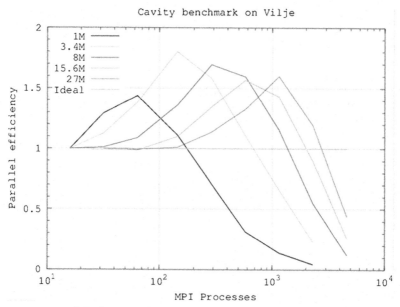

Figure 6.14 Parallel efficiency curves for different problem sizes. Notice that the ideal curve has 100% efficiency.

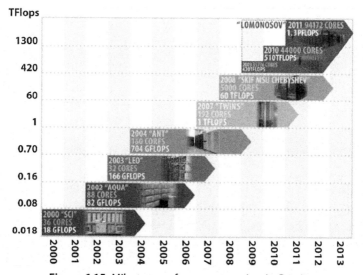

Figure 6.15 Milestones of supercomputing in Russia.

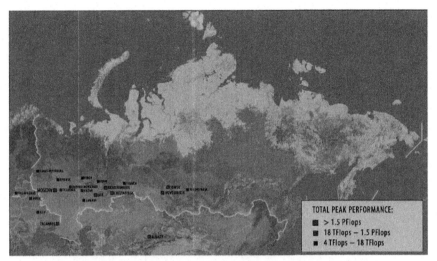

Figure 6.16 Places in Russia where clusters have been deployed.

cores (depending on the problem size and other overheads), just as the speedup drops after a certain number of cores.

5. CASE STUDIES OF SOME RENOWNED CLUSTERS

5.1 t-Platforms' Lomonosov HPC

t-Platform is a leading company from Russia in the field of HPC. Germany, the United States, and China are the main competitors in the field of HPC, but the Russians are also on their way to take the lead. This became clear to the world in November 2009, when Lomonosov became the 12th in the Top500 Web site list of the world's largest supercomputers. Figure 6.15 shows the history of supercomputing in Russia. The graph shows the increment in teraflops achieved in the past decade of supercomputing at Moscow State University (MSU). The last three years show that HPC machines in Russia have broken the barrier of teraflops per second by entering into the petaflop range. A number of supercomputing centers working in Russia are shown in Figure 6.16 [3].

5.2 Lomonosov

The Lomonosov is the largest supercomputer in Moscow, Russia, located at MSU. It was named after renowned Russian scientist M.V. Lomonosov [3]. The peak performance of Lomonosov is 420 teraflops and LINPACK

Figure 6.17 Lomonosov cluster with inset of M.V. Lomonosov.

Figure 6.18 Water tanks installed in the basement to cool the gigantic cluster.

measured performance is 350 teraflops. System efficiency is 83%, which is considered the best in the world in terms of the performance of super-computers. The Lomonosov is based on the T–Blade 1.1i and T–Blade 2T and TB2–TL, which are equipped with GP–GPU nodes. All of the system is in-house except the processors, which are Intel or NVIDIA Tesla, power supply units, and cooling fans. Figure 6.17 shows the hall in which the Lomonosov cluster was installed.

5.2.1 Engineering Infrastructure

The Lomonosov cluster consumes around 1.36 MW of power and has redundant power supply units in case of failure. The uninterruptible power supply (UPS) system is guaranteed to provide sufficient power and cool down the system for the time required to shut down running tasks gracefully and shut the system down appropriately. Two UPS units provide separate power to the two segments of Lomonosov, each with a performance of 200 teraflops.

In addition, in case of power loss, one compute segment can be powered down to allocate more power to the critical computational segment. The specialty of the UPS system is 97%, which is above the conventional 92% used at the industrial level. High efficiency is a must for such huge computational systems.

Another feature of Lomonosov that makes it ubiquitous is its high-scale computational density workload, which draws 65 kW of power per a 42-U, approximately 73.5 in rack. A separate cooling system (Figure 6.18) that occupies an 800 m^2 room provides cooling for this massive structure. Because of the long winters in Russia, the system is also cooled by the free outside atmosphere, by cutting off compressors running through water chillers. This helps reduce power consumption for about half the year. The system is also equipped with a fire security system. Within a half second the automatic fire system fills the entire room with a gas, terminating fire without damaging any of the equipment components. The fire is suppressed but it does not lower the concentration of oxygen in the room, and thus it is relatively safe for personnel.

5.2.2 Infrastructure Facts

Preparation for this gigantic mega-structure involves reinforcing floors to accommodate rack cabinets weighing more than 1100 kg each, as well as insulating the data center walls to keep a nominal 50% humidity. Six water tanks are used for the water circulation system, carrying over 31 tons of water to provide necessary cooling. The UPS and the cooling and management subsystems are tightly coupled. The system has a two-stage scenario in which it analyzes the first 3 min of an event to determine whether the power loss was temporary, in which case normal operation can be restored, or whether it is a permanent situation, in which case the system will restart a proper shutdown procedure, creating backups for all running jobs. In the event of external power loss, the entire system takes 10 min to shut down completely. The cooling system consists of an innovative hot–cold air containment system; high-velocity air outlets provide efficient, even air mixed with minimal temperature deviations in the hot aisle zones.

Table 6.14 Key features of the Lomonosov cluster

Features	Values
Peak performance	420 teraflops/s (heterogeneous nodes)
Real performance	350 teraflops/s
LINPACK efficiency	83%
Number of compute nodes	4446
Number of processors	8892
Number of processor cores	35,776
Primary compute nodes	T-Blade 2
Secondary compute nodes	T-Blade 1.1 peak cell S
Processor type of primary compute node	Intel Xeon X5570
Processor type of secondary compute node	Intel Xeon X5570, power cell 8i
Total RAM installed	56 TB
Primary interconnect	QDR Infiniband
Secondary interconnect	10G ethernet, gigabit ethernet
External storage	Up to 1350 TB, t-Platforms ready storage SAN 7998—Lustre
Operating system	ClusterX t-Platforms edition
Total covered area occupied by system	252 m^2
Power consumption	1.36 MW

5.2.3 Key Features

Key features of the Lomonosov cluster are given in Table 6.14. The details consider only the Intel Xeon Westmere series and not NVIDIA GPUs.

5.3 Benchmarking on TB-1.1 Blades

This benchmarking was performed on TB-1.1 blades. These blades are made by MSU HPC team and contained 264 cores in 16 blade enclosures. Thus, each blade contained two processors. The tests were performed in two stages. In the first stage the processors used were AMD Opteron 6174 (Magny-Cour), each with 12 cores, and so 24 cores per blade. In the second stage the processors used were Intel Westmere X5670, each with a hex core, and so 12 cores in a single blade. The performance curves were tested for 3 million and 8 million cells of a CFD problem. The problem consisted of an NACA 0012 aerofoil. These curves were tested for 256 cores of an AMD Magny-Cour processor and a 128 Intel Xeon Westmere processor. Figure 6.19 shows the curve for the 3-million cell problem in ANSYS Fluent and Figure 6.21 shows a plot of the 8-million cell problem in

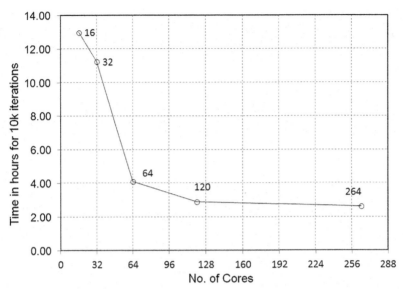

Figure 6.19 Performance results of 3-million mesh size problem run on an Intel Westmare hex core processor.

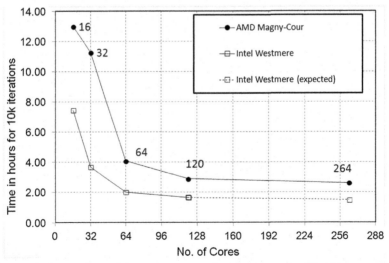

Figure 6.20 Comparison of the two processors' performance for a 3-million mesh size problem.

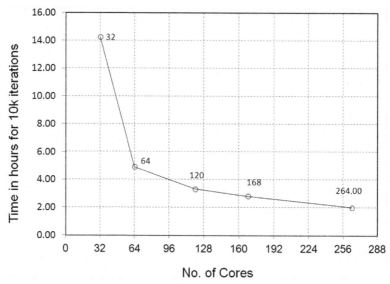

Figure 6.21 Performance results of an 8-million mesh size problem run on an Intel Westmare hex core processor.

ANSYS Fluent performed with AMD Opteron 6174 (Magny-Cour) processors. Figure 6.20 and Figure 6.22 show the performance curves of Intel Westmere X5670 and a comparison with AMD cores.

This shows that Intel performs much faster than an AMD processor. An expert from t-Platforms discussed this and said that Intel Xeon performance is much better with AMD Magny-Cour. This is because of the overall clock speed of Intel and AMD. The AMD Magny is 2.2 GHz whereas the Intel Xeon is 2.93 GHz. Overall, for both AMD and Intel it is obvious that for a single problem, as the number of cores increases the time to complete 10,000 iterations decreases until it show no significant decrease with increasing cores. The IST team is thankful to t-Platform for this ultimate support in benchmarking.

5.4 Other t-Platform Clusters

5.4.1 Chebyshev Cluster

The MSU Chebyshev supercomputer (Figure 6.23) with 60 teraflops/s of peak performance was the most powerful computing system in Russia and eastern European countries before Lomonosov. It was named after the Russian mathematician P.L. Chebyshev. The Chebyshev, with 60 tera-flops/s of peak performance, was based on 625 blades designed by

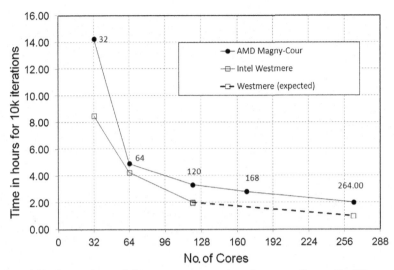

Figure 6.22 Comparison of the two processors' performance for an 8-million mesh size problem.

Figure 6.23 Chebyshev cluster.

t-Platform, incorporating 1250 quad-core Intel Xeon E5472 processors. The LINPACK real performance test result was 47.17 teraflops/s or 78.6% peak performance. The MSU Chebyshev supercomputer incorporated the most recent technological findings from the industry and used several in-house developed technologies. Its computing core used the first Russian-developed blade systems, which incorporate 20 quad-core Intel Xeon 3.0 GHz 45 nm processors in a 5-U chassis, providing the highest computing density among all Intel-based blade solutions on the market.

5.4.2 SKIF Ural Supercomputer

The SKIF Ural Supercomputer incorporates advanced technical solutions. It incorporates over 300 up-to-date, 45 nm Intel Hypertown quad-core processors. Along with the SKIF MSU supercomputer, SKIF Ural was the first Russian supercomputer of the time built using the Intel Xeon E5472 processors. This supercomputer is also equipped with advanced engineering software for research involving analysis and modeling. FlowVision is an example and is Russia's indigenous computer product. There are many application areas, the foremost of which is CFD. Others include nanotechnology, optics, deformation and fracture mechanics, 3D modeling and design, and large database processing. In March 2008, SKIF Ural occupied the fourth position in the eighth edition of the Top50 list of the fastest computers in the Commonwealth of Independent States countries.

5.4.3 Shaheen Cluster

The Shaheen cluster consists of a 16-rack IBM Blue Gene/P supercomputer owned and operated by King Abdullah University of Science and Technology (KAUST). Built in partnership with IBM, Shaheen is intended to enable KAUST faculty and partners to research both large- and small-scale projects, from intuition to realization. Shaheen is the largest and most powerful supercomputer in the Middle East. Originally built at IBM's Thomas J. Watson Research Center in Yorktown Heights, New York, Shaheen was moved to KAUST in mid-2009 [4]. The creator of Shaheen is Majid Alghaslan, KAUST's founding interim chief information officer and the university's leader in the acquisition, design, and development of the Shaheen supercomputer. Majid was part of the executive founding team for the university and the person who named the machine.

5.4.3.1 System Configuration

Shaheen includes 16 racks of Blue Gene/P, with a peak performance of 222 teraflops. It also contains 128 IBM System X3550 Xeon nodes, with a peak performance of 12 teraflops. The supercomputer contains 65,536 independent processing cores. The Blue Gene/P is technology evolved from Blue Gene/L. Each Blue Gene/P motherboard chip contains four PowerPC 450 processor cores running at 850 MHz. The cores are cache coherent and the chip can operate as a four-way symmetric multiprocessor. The memory subsystem on the chip consists of small private L2 caches, a central shared 8 MB L3 cache, and dual DDR2 memory controllers. The chip also integrates the logic for node-to-node communication, using the same network topologies as Blue Gene/L, but at more than twice the bandwidth. A compute card contains a Blue Gene/P chip with 4 or 4 GB DRAM, comprising a compute node. A single compute node has a peak performance of 13.6 gigaflops. Thirty-two compute cards are plugged into an air-cooled node board. A rack contains 32 node boards (thus, 1024 nodes and 4096 processor cores).

6. CONCLUSION

This discussion shows how various supercomputer manufacturers market their product by benchmarking. It gives the idea that if users want to make their own cluster or want to hire some vendor, they can easily obtain their desired machine by analyzing standard benchmark curves. However, users must know mesh requirements beforehand. Also, only the speedup curves are important, but also the efficiency, as shown in tabular form for ANSYS Fluent and CFX and in graphical form for OpenFOAM. Budgeting is also important; it is not advisable to buy an expensive machine (such as IBM) if your problems usually run on less than 250 cores. IBM machines are useful for big problem sizes of more than 10 million. This chapter had no rocket science behind it; it was an information guide for establishing an economical and productive HPC cluster.

REFERENCES

[1] William Aiken. Sun business ready HPC for ANSYS FLUENT-Configuration guidelines for optimizing ANSYS FLUENT performance. ISV engineering, sun BluePrints™ online, part No 821-0696-10, Revision 1.0, September 4, 2009.
[2] ANSYS HPC Benchmark, http://www.ansys.com/benchmarks, visited September 2014.
[3] Moscow State University and High Performance Computing, presentation Vladimi Voevodin, deputy director, Research Computing Center from Moscow State University, Helsinki, Finland – April 13, 2011.
[4] Webpage on, Michael Feldman, HPC Wire, Saudi Arabia Buys Some Big Iron, October 1, 2008. Retrieved 2009-03-13. Accessed on 28th March 2015.

CHAPTER 7

Networking and Remote Access

1. TRANSMISSION CONTROL PROTOCOL/INTERNET PROTOCOL

When you send a letter to your friend living abroad, you write more than his name in the "send to" part: You also write his complete address. In the address you say more than just the city name, like London or Manchester, you write the town name, the street address, and the house number. In the same way, when you access the network, your domain name server Internet protocol (IP) is a particular address that you specify in the Internet explorer. This particular address, which you usually do not see, is the IP. It is a unique number consisting of a set of four three-digit numbers separated by a dot. What you usually see is the Web address, such as http://www.google.com or http://www.yahoo.co.uk. Before we proceed further, it seems worth mentioning that this portion could be boring to a computational flow dynamics (CFD) user, because many things are related to information technology or computer sciences, but they are equally linked to CFD jobs because networking is the backbone of high-performance computing (HPC).

With this understanding of the importance of an IP address, we will go into further into its details. There must be some standard way to address the IP; otherwise, data can be lost as in the case of mail delivery when you give no address or an ambiguous one. Standardization is done by the Internet Assigned Numbers Authority, which assign IPs to the international domain and to organizations. Small organizations or individual users can obtain an IP from Internet service providers.

Internet protocol addressing is done by assigning a 32-bit address code, which is a set of four three-digit numbers. The address in binary number format may look like:

11000011 00100010 00001100 00000111

which in decimal format can be written as:

195.34.12.7

This is much easier to remember than the binary format. The type of number formatting is further divided into two parts: the network and the host node on the station of the network.

Using HPC for Computational Fluid Dynamics
ISBN 978-0-12-801567-4
157

1.1 Internet Protocol Classes

Based on this division into two parts, IP addresses involve different types of classes. There are five standard classes of IP addresses. They are constructed using different combinations of node and host, as shown in Figure 7.1. The classes are categorized as A, B, C, D, and E. Figure 7.2 explains this in detail for each class:

- **Class A** Class A addresses use an 8-bit network number and a 24-bit node number. Therefore, Class A addresses can have up to $2^{24} - 2 = 16,777,214$ hosts on a single network. Class A addresses are in the range: 10.0.0.0 to 127.255.255.255.
- **Class B** Class B addresses use a 16-bit network number and a 16-bit node number. Thus Class B addresses can have up to $2^{16} - 2 = 65,534$ hosts on a network. Class B addresses are in the range (beginning with 128 because 127 belongs to Class A addresses): 128.1.0.0 to 191.255.255.255.
- **Class C** Class C addresses use a 24-bit network number and an 8-bit node number. Class C addresses can have up to $2^8 - 2 = 254$ hosts on a network. Class C addresses are in the range (beginning with 192): 192.0.0.0 to 223.255.255.255.
- **Class D** Class D addresses are used for multi-casts (messages sent to many hosts). Class D addresses are in the range: 224.0.0.0 to 239.255.255.255.
- **Class E** These addresses are for experimental use.

The address with all zeros in the host address (for example, in class C, 192.168.0.0 (11000000.10101000.00000000.00000000)) is the network address and cannot be assigned to any machine. The address with all ones in the host address (11000000.10101000.00000000.11111111 = 192.168.0.255) is

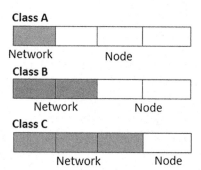

Figure 7.1 Internet protocol class distribution. Grayed portions are for the network host.

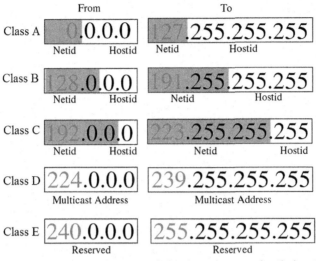

Figure 7.2 Internet protocol distribution: more detailed.

the network broadcast address. This addressing structure allows IP addresses to uniquely identify each physical network and each node on each physical network. For a particular value of the address, the base address of the host with all zeros is known as the network address and not allotted to the host. Similarly, the top address of the range host address of all ones is not assigned because it is used for broadcasting data to all hosts with the same network address.

1.2 Special IP Addresses

Some special types of IP addresses are unique and are used for a particular type of work:

- Network address: All of the host identifications (IDs) in this type of address contain all zeros (in binary notation).
- Direct broadcast address: In this address, all of the host IDs contain all ones. The direct broadcast address is used by the router to send a message to every host on a local network. Every host/router then receives and processes the data with direct broadband address.
- Limited broadcast address: The net ID and host ID contain all ones (in binary notation). A limited broadcast address is used by a host to send a packet to every host on the same network. The data packets are blocked by the router to confine the packets to the local network.
- Loop back address: This kind of address begins with 127, so it is always of the type 127.x.y.z. In FLUENT, when accessing multiple personal

computers (PCs), it identifies the nodes on the basis of their IPs. This creates problems if the file/etc./hosts contain a loop back address at the end that is not commented by default. To run FLUENT that this line must be commented.

1.3 Subnetting

Consider a network with millions of hosts. For example, class A contains over 16 million host IDs, which is not practical. This also causes broadcast problem and IP addresses are not used efficiently in this way. A remedy for this is to make groups in the host IDs so that they are used effectively. It also reduces network traffic and makes it simple to manage, and contains smaller broadcast domains. Subnetting an IP network is done for a variety of reasons, including using different physical media (such as Ethernet, FDDI, and WAN), preserving address space, and reasons of security. The most common reason is to control network traffic. In an Ethernet network, all nodes on a segment see of all the packets transmitted by all other nodes on that segment. Performance can be adversely affected under heavy traffic loads, owing to collisions and resulting retransmissions. A router is used to connect IP networks to minimize the amount of traffic each segment must receive. In subnetting, large networks need to be divided into smaller networks. These smaller divisions are called subnetworks and provide addressing flexibility. Most of the time subnetworks are simply referred to as subnets. A primary reason for using subnets is to reduce the size of a broadcast domain. When broadcast traffic begins to consume too much of the available bandwidth, network administrators may choose to reduce the size of the broadcast domain.

Similar to the host number portion of class A, B, and C addresses, subnet addresses are assigned locally, usually by the network administrator. Like other IP addresses, each subnet address is unique (Figure 7.3).

Thus, the subnet number simply gets some bits from host and helps the network to use them. It determines which part of an IP address is the network field and which is the host field. It can be 32 bits long and has 4 octets, just like an IP address.

Figure 7.3 The Subnetting.

1.4 Subnet Masking and Example

Applying a subnet mask to an IP address allows you to identify the network and node parts of the address. The network bits are represented by the ones in the mask and the host bits are represented by the zeros. Performing a bit-wise logical AND operation between the IP address and the subnet mask results in the network address or number. A class B address can be effectively translated. For example, the IP address of 172.16.0.0 is assigned, but host addresses are limited to 255 maximum, allowing eight extra bits to use as a subnet address. The IP address of 172.16.97.235 would be interpreted as IP network address 172.16, subnet number 97, and node number 235. In addition to extending the number of addresses available, subnet addressing provides other benefits. It allows a network administrator to construct an address scheme for the network by using different subnets for other remote locations in the network or for other departments in an organization. The previous example uses the entire third octet for a subnet address, but this is not a restriction. To create more network numbers, just shift some bits from the host address to the network address. For instance, to partition a class C network number 192.68.135.0 (apologies if someone has same IP address) into two, you shift 1 bit from the host address to the network address. The steps are as follows:

1. Write down the network number in binary form, such as 192.68.135.0 (11000000.01000100.10000111.00000000).
2. Write down the net mask as 255.255.255.128 (11111111.11111111.11111111.10000000) (128 is just an example).
3. Now add these binary with logical and i.e., $1 + 1 = 1$ and $1 + 0 = 0$, $(11000000 \cdot 01000100 \cdot 10000111 \cdot 10000000) = 192.68.135.128$.
4. The subnet has network number 192.68.135.0 with hosts 192.68.135.1 to 129.68.135.126
5. The address 192.68.135.127 is not assigned because it is the broadcast address of the subnet.

Similarly, if the subnet mask contains the last digit as 240, it would mean that the binary notation would be 11110000. In this way, the network can be split into 14 subnets of 14 nodes each. This will limit the network to 196 nodes instead of the 254 it would have had without subnetting, but it has the advantages of traffic isolation and security. To accomplish this, a subnet mask 4 bits long would be used. Table 7.1 explains this.

Table 7.1 Subnetting with 14 possible combinations

Subnet bits	Network number	Node addresses	Broadcast address
0000	192.68.135.0	Reserved	None
0001	192.68.135.16	0.17−0.30	192.68.135.31
0010	192.68.135.32	0.33−0.46	192.68.135.47
0011	192.68.135.48	0.49−0.62	192.68.135.63
0100	192.68.135.64	0.65−0.78	192.68.135.79
0101	192.68.135.80	0.81−0.94	192.68.135.95
0110	192.68.135.96	0.97−0.110	192.68.135.111
0111	192.68.135.112	0.113−0.126	192.68.135.127
1000	192.68.135.128	0.129−0.142	192.68.135.143
1001	192.68.135.144	0.145−0.158	192.68.135.159
1010	192.68.135.160	0.161−0.174	192.68.135.175
1011	192.68.135.176	0.177−0.190	192.68.135.191
1100	192.68.135.192	0.193−0.206	192.68.135.207
1101	192.68.135.208	0.209−0.222	192.68.135.223
1110	192.68.135.224	0.225−0.238	192.68.135.239
1111	192.68.135.240	Reserved	None

2. SECURE SHELL

With many Internet protocols such as Telnet, File Transfer Protocol (FTP), and Remote SHell (RSH), a problem was that while accessing the remote servers, the passwords were also transmitted in plain text, which was highly insecure. Common use of .rhost is an example. Thus, while connecting to the outside world domain through the Internet, a secure system of logging on was devised. Secure SHell (SSH) was produced for this purpose. It was developed by SSH Communications Security, Ltd and has both Linux- and Windows-based versions available. In Red Hat Linux, however, you do not need it.

Install SSH explicitly because the package offers installation. Some repositories for additional functionality may be obtained from the Internet. Secure SHell encodes all communications between two end points, eliminating the chance that passwords or other sensitive bits of information are discovered by intermediate sniffers (hackers). Secure SHell uses a public-key authentication-based system that is required for login into the remote server after you provide the password. For this purpose, RSH is now obsolete because as an authentication model it was insecure in a cluster environment. Secure Copy (SCP) is also part of safe and sound file copying between the two nodes of a cluster. OpenSSH is becoming popular for Linux-based systems because it is well-supported, easy to install, and portable.

2.1 Setting Name Resolution

Sometimes when you try to access remote nodes using their names, it displays a message about temporary failure in name resolution. This is because that the SSH does not know to which IP the remote node belongs. This can be easily solved by mentioning host names in the /etc/ hosts file. If you do not yet have access to nodes through SSH, it would contain only one IP address, 127.0.0.1 which is a loop back and should be commented. In ANSYS FLUENT it does not access remote nodes because it searches for the loop back address. Thus, first mention the IP of the host PC (which you are currently on) and then its name, and then the domain name.

Similarly, in the second line mention the PC which you want to access, and then mention its IP, its name, and then the domain name. Now you are ready to log in through SSH. It is better to log out the terminal or open up a new terminal for SSH remote nodes.

If you are on node01, type "ssh node02" where node02 is the name of the remote node you want to access. The first time, it will generate fingerprints and then ask to connect (yes or no?). Type "yes" and then it will ask for password. Type the password and it will log into the remote machine. To exit the node, type "exit."

2.2 Setting a Password-Less Environment

While accessing the remote node, SSH asks for fingerprints only for the first time, as shown below, and stores the information into the /root/.ssh/ folder.

```
[root@node1] ssh-keygen -t rsa.
Generating public/private rsa key pair.
Enter file in which to save the key (/root/.ssh/id rsa):
Enter passphrase (empty for no passphrase):
Enter same passphrase again:
Your identification has been saved in/root/.ssh/id rsa.
Your public key has been saved in/root/.ssh/id rsa.pub.
The key fingerprint is:
f6:82:g8:27:35:cf:4c:6d:13:22:30:cf:5c:c8:a0:23
```

After this, whenever you log into the other node it will ask only for a password. Now, if we talk about ANSYS FLUENT parallel in an SSH environment, there must be no such thing as a password because Fluent cannot provide one. It would only pick up the local processors of the host PC. Thus, a password-less environment is mandatory.

For SSH, one must remember the two important file names "autho-rized keys" and "id rsa.pub". They both are in the /root/.ssh/ folder. "id rsa.pub" contains the public key of the current PC you are logged into. The file "authorized keys" contains the list of the public keys of all of the remote PCs to which you want to connect and the host PC itself. Initially in the .SSH directory you will have **No** file in it. Probably one file would exist if you logged in multiple times to your PC or remote PC. You do not need to worry about that. However, from anywhere (apart from root) you can start generating the files in the .SSH folder. To generate the files, type the following (assuming this computer name is node1):

```
[root@node1] ssh-keygen -t rsa.
```

This will create the id_rsa and id_rsa.pub in the folder /root/.ssh. id_rsa will contain the private key while id_rsa.pub will contain the public key. You need to copy this to your own authorized keys file as well as to the authorized keys file of other PCs. To avoid error from manual copying, type the following command:

```
[root@node1] cat/root/.ssh/id rsa.pub | ssh root@node2 'cat
>>/root/.ssh/authorized keys'
```

Note: the above statement must be typed in one line on the terminal; because of brevity, the page in this chapter was typed on two lines.

This will copy the keys into the authorized keys of node1 into the authorized keys file of node2.

2.3 Remote Access on a Linux Platform

Networking acts as a backbone in clustering. Not having networking is the same as constructing a building without pillars. The network allows the nodes to communicate and exchange data. An excellent network is one that has low latency and high bandwidth.

Common networks in use today are Gigabit and Infiniband. They differ mainly with respect to the speed they provide. When you are ready to submit your job there is one more intermediate task.

Clusters work in a batch environment. The basic idea is that you create a task and make a script to submit the task or job. This job is submitted via a job scheduler. As you submit a job, if everything goes fine your job will start running. It may also wait in a queue, but that is not an issue because it may not have enough resources to run at the time you submit the job. Be patient. Within the script you tell the cluster many things, normally including your job name, the number of cores, the number of nodes, your e-mail address, the time to wait in a queue, the time to running the job, etc. A typical script for ANSYS FLUENT is mentioned below with PBS job scheduler.

```
#PBS -N name
#PBS -l walltime=10:00:00
#PBS -l nodes=2:ppn=2
#PBS -j oe
# Run FLUENT
fluent 2ddp -t4 -cnf=$PBS NODEFILE -g -i input.txt > logfile.$$
```

2.3.1 PuTTY

Most of the cluster environments are Linux-based, so one way to access the cluster via your remote desktop Windows-based PC is through PuTTY. PuTTY is an SSH client for remote access. This requires the IP of the cluster control node and its username and password (Figure 7.4). After you log into the PC for the first time, it asks to store the information you provided into the cache, similar to the login procedure when you do it for the first time for SSH (Figure 7.5). Then you enter the username and password and log in. You will see a Linux terminal opened (usually black).

Now you are ready to submit your job if your script is ready. To submit the job, a job scheduler is necessary. Even if you do not have a job scheduler, you can successfully run your tasks of ANSYS FLUENT; when you do not have a time limitation and not many users to work on a cluster, so you can easily work without a job scheduler. There are three options for running your task: (1) without a job scheduler or graphics; (2) without a job scheduler and with graphics; and (3) with a job scheduler.

2.3.1.1 Without a Job Scheduler or Graphics

If you want to run your job of Fluent without a job scheduler, it is useful to set up your case and data files on your own PC before running it onto the cluster. Then, transfer the files onto the cluster account via WinSCP and after that open up the PuTTy terminal and log in through your username and password. Then launch the ANSYS FLUENT via the command as follows:

```
fluent 3ddp -t16 -ssh -g -i <input> output 2&>1.
```

This will run a 3D double-precision (dp) version of Fluent using 16 cores (t16) without graphics (-g) and through ssh (-ssh) using the input file (input.txt)and two& runs the process in background. After Fluent launches, you will see something strange: If you have all of the FLUENT licenses

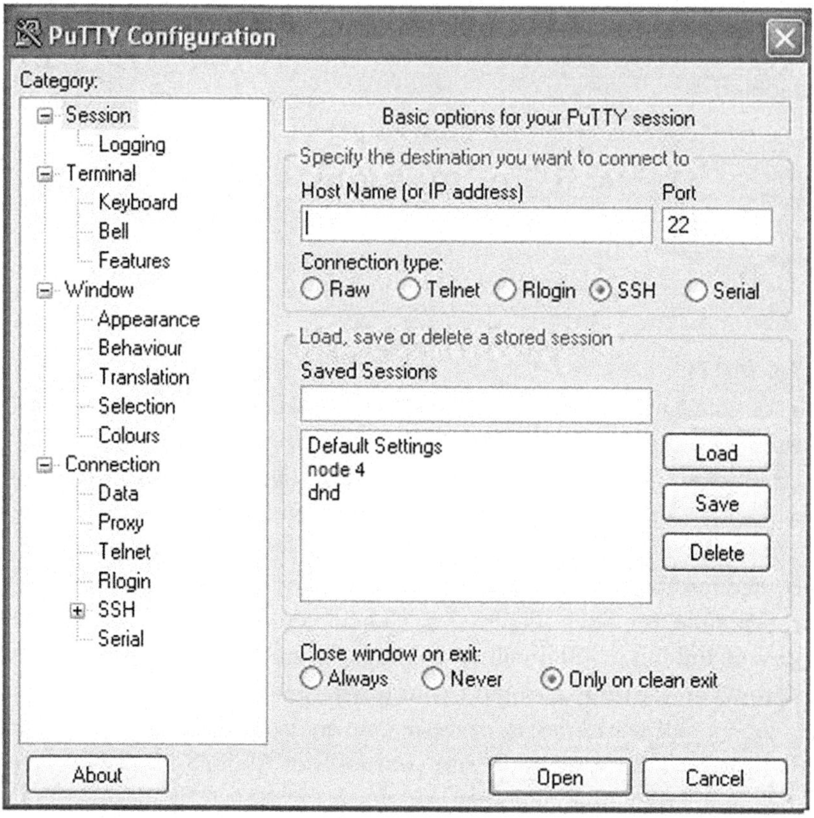

Figure 7.4 The PuTTY console.

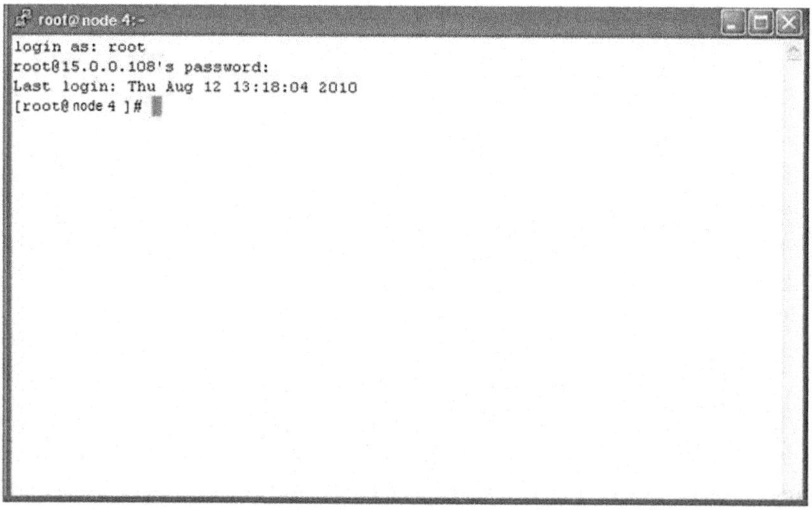

Figure 7.5 The PuTTY Linux terminal.

available, it will pick the 16 cores of only your current node. No other machines in the list of the machines will be shown by FLUENT. Why is it so? It is because you have not told ANSYS FLUENT to pick up the machine file containing machine names.

For this purpose, type the following:

```
fluent 3ddp -t16 -ssh -g -i -cnf=host <input> output 2&>1.
```

where the host file contains the name of machines to be used. The correct path must be given if the file is somewhere other than the FLUENT directory. The following must be taken into account while writing the host file. Type the host name first. If the machine is multi-core and you have all the licenses available, write the machine name as many times as the number of cores. Thus, for a quad-core machine you will type the machine name four times. After entering all of the machine names, leave one extra line by pressing Enter and then save and quit your file. Do all of this in the PuTTy or Linux terminal through the vi editor. Caution: Do not edit any text document like that in Windows because the format will change and Linux will not recognize it. The input file will contain the commands to be executed by Fluent after running the above command. This is not Linux-specific; rather, it is FLUENT specific. A sample is shown below:

```
rc mytask.cas.
rd mytask.dat.
it 1000 wc mytask-1000.cas.
wd mytask-1000.dat.
exit
```

This will read case and data files, iterate for 1000 iterations, and, if everything goes fine, write case and data files. A number of commands for different tasks in Fluent can be found in the Fluent Text Command List. You may also write your output to an output file that will show the output of all of your iterations as you Fluent-finish the calculations.

2.3.1.2 Without Job Scheduler and with Graphics

The method is not different from the one described above. An exception is that a third piece of software, Xming, is used for graphics options. You need to install Xming on your local Windows PC. While in the PuTTy console you will put check the SSH tab to enable the X11 forwarding option. When you will launch PuTTy and you type a command such as gedit on the terminal, it will open the gedit window. This will not work if Xming is not installed and you have not enabled X11 forwarding.

Similarly, when you type "Fluent," Fluent will run in graphics mode with its GUI. If you have a low-bandwidth connection between your PC and the remote cluster, it could be a bottleneck because the graphics will take a lot of time to load and it will take many seconds to respond after you click. When launching Fluent, you type the same command as mentioned above and drop the -g switch. If you are far behind the cluster in another city or country, you will run your work through the Internet. You will log in using ssh -X <username>@<domain name> and the -X will enable the graphics option in Linux mode.

2.3.1.3 With Job Scheduler

In the job scheduler you do not see the graphics. You can do so by saving the outputs momentarily so that you can view them using another PuTTy console. The job scheduler helps you manage your work and sets your job in a queuing system. If some nodes are free as you requested, your job will start immediately. The job submission command is qsub <scriptname>. To check the status of the job, type "qstat." A typical script for the job submission of a Fluent task is given below:

```
#PBS -N name.
#PBS -l walltime=10:00:00
#PBS -l nodes=2:ppn=2
#PBS -j oe
# Run FLUENT
fluent 2ddp -t4 -cnf=$PBS NODEFILE -g -i input.txt > output.$$
```

The first line should contain the job name. The second line specifies the running time of the job. The third line specifies the number of nodes followed by the term "ppn," which means processor (cores) per node. This can be used intelligently: for example, you have five nodes and you want to run your job on 24 cores, so you should not run you job on five nodes, but rather use three nodes and run with eight cores per node (assuming eight cores per node). The last line specifies the job output name. All of the lines with a hash symbol followed by "PBS" are job scheduler commands; the rest are comments. The last line of the job scheduler contains the command for FLUENT execution.

The explanation given above is for accessing the local area network via a single domain. However, in practice it is different. When you are sitting miles from your cluster, you normally gain access through the Internet. From PuTTy you first log into a login node and then from the login node you access the nodes as described above. If you gain access through the Internet, first type the username and password of the login node. Then type:

```
ssh -X username@<; IP address>
```

where the username would be your username and the IP address would be the IP of the login node network or the domain name server. Then you can SSH your other nodes in the usual manner. The −X switch enables the graphics option in accessing SSH. Consequently, writing −x instead of −X would disable it.

3. WINSCP

WinSCP stands for Windows Secure Copy, a free and open-source SFTP and FTP client for Microsoft Windows. Its main function is secure file transfer between a local and a remote computer. Beyond this, WinSCP offers basic file manager and file synchronization functionality. For secure transfers, it uses Secure Shell (SSH) and supports the SCP protocol. When you want to transfer data from your user node on Windows to the control node, you transfer files through this software. This has a much more user-friendly layout than PuTTY. WinSCP contains two panels: the left is the user computer (which is Windows-based) and the right is a Linux-based control node, as shown in Figure 7.6.

4. RUNNING ANSYS FLUENT IN AN WINHPC ENVIRONMENT

Microsoft introduced Windows HPC Server 2008 to run HPC applications in a user-friendly environment. Because Linux is dry and requires memorizing a lot of commands, WinHPC lets the user do all of it in a GUI environment. However, in my opinion (most readers might disagree with me) Windows cannot compete with Linux. Ninety percent of the clusters in the world run on Linux. However, Tianhe-1A has the provision of performing calculations using WinHPC 2008. This is because Windows has certain issues that do not exist with Linux. This is a separate discussion; the main focus here is to explain how to run ANSYS FLUENT in an WinHPC environment. Images of ANSYS FLUENT menus are given for a clear understanding of the process.

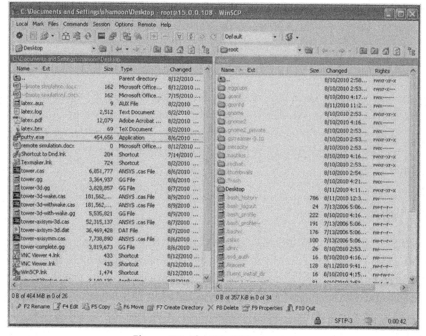

Figure 7.6 The WinSCP console.

4.1 ANSYS FLUENT Installation with a WinHPC Platform

1. Install and configure ANSYS FLUENT network parallel HPC

    Before installing FLUENT, we assume that Microsoft HPC is installed and configured properly and that the compute nodes can access the head node. The Windows HPC Server 2008 Guide can be accessed to solve WinHPC issues, at http://technet.microsoft.com/en-us/library/cc793950.aspx. This guide is also available with the installation files for Microsoft HPC Pack 2008 (HPC GettingStarted.rtf, in the root folder).

2. Install ANSYS FLUENT (it is only necessary to install FLUENT on the head node).

    The new directory structure will install FLUENT in:

    `C:\Program Files\ANSYS Inc\v1xx\fluent`.

3. Share the Fluent directory that sits under `C:\Program Files\ANSYS Inc\v1xx`, where xx is your ANSYS version number, so that all computers on the cluster can access this shared directory through the network (Figure 7.7).

Figure 7.7 The ANSYS Fluent path.

4. Setup your ANSYS FLUENT working directory as a shared network drive. (A working directory is the directory where your case and data files reside.)

5. Go to **Start Menu**, **Computer**, and select **Map Network Drive** from the menu near the top of the screen (Figure 7.8).

6. Select a Drive letter and then press **Browse** ... to your working directory: for example, C:\Working\ (Figure 7.9).

7. Configure HPC clients

   A client machine is any machine on the network that can access the cluster through the network. The following requirements are prerequisite.

   a. Client machines must run Microsoft Windows XP 64-bit, Windows Vista 64-bit, Windows 7 64-bit, or Microsoft 2008 Server 64-bit.

   b. If you only have 32-bit clients, you can run FLUENT on the HPC cluster but you must run in batch mode using a journal file.

   c. Client machines must have a high-end video card with the latest graphics driver preinstalled by the vendor.

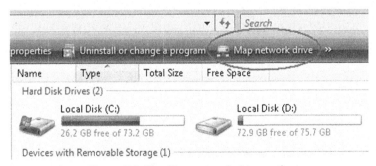

Figure 7.8 The Map network drive path.

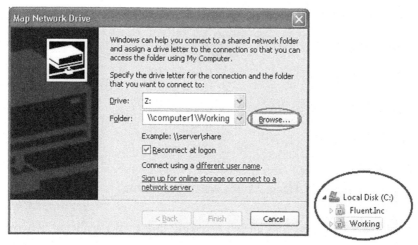

Figure 7.9 Mapping network drive in Windows.

8. Remote Desktop Connection support

Many times, launching ANSYS FLUENT using Remote Desktop Connection will not result in display issues if the HPC cluster head node has a PCI Express ×16 Graphics card installed with the latest graphics driver and if the client machines are using the latest Remote Desktop Client (version 6). If these conditions are not met, you must launch FLUENT from a 64-bit client machine after installing Microsoft HPC Pack.

9. ANSYS FLUENT startup command

If you are running via Remote Desktop Connection, the default driver FLUENT will use is Microsoft Windows. If you run into graphics issues, check which driver ANSYS FLUENT is using.

- Open up FLUENT
- Choose Help menu, Version
- Check Graphics Version. If it reads msw/win, you are using the operating system Windows driver.

```
Cortex Version: 3.9.1
Graphics Version: 14.17-1, msw/win
OS Version: Windows NT 6.0.6001
```

Figure 7.10 Windows graphic card information message.

- Try using the OpenGL driver by starting FLUENT with the following flag: fluent -fluent options "-driver opengl" as shown in Figure 7.10.
- Check Graphics Version once FLUENT is launched, to verify that it is using the OpenGL driver.

10. Configuring HPC client machines to access the cluster

   a. Install HPC Pack 2008 on client machines from the head node network share called `C:\Program Files\Microsoft HPC Pack\Data\reminst` or from the CD.

   b. Open up the reminst directory and double-click on the setup.exe file. Some additional programs may be required: for example, .NET Framework. The installer will prompt you to install these programs.

11. Options with FLUENT

   a. It is easiest to make a shortcut to the fluent.exe file from the head node onto the client machines. The fluent.exe is usually located in `C:\Program Files\ANSYS Inc\v1xx\fluent\ntbin\win64`.

   b. Launch FLUENT from the shortcut on your desktop.

   c. Choose your Dimension, Display Options. Under Options, choose Double Precision if necessary. Choose Use Microsoft Job Scheduler. You can also request resources but delay the launching of ANSYS FLUENT until the actual resources are allocated. In the ANSYS FLUENT GUI, choose the **Scheduler tab** then check **"Start When Resources are Available."**

   d. Under Processing Options, choose Parallel per MS Job Scheduler and then enter the number of processes you will be using.

   e. Select **Show More.** The menu will expand as shown in Figure 7.11; the expanded menu is shown in Figure 7.12.

   f. Select the **Parallel Settings** tab, as shown in Figure 7.13.

   g. Select the appropriate options under Interconnects and MPI Types (Ethernet msmpi are the defaults when running on 2008 HPC Server).

   h. Choose the Scheduler tab and type in the name of the head node (Figure 7.14).

   i. If you will be compiling and loading User Defined Functions (UDFs), choose the UDF Compiler tab and check the box Setup Compilation Environment for UDF, as shown in Figure 7.15.

   j. Press OK to launch ANSYS FLUENT.

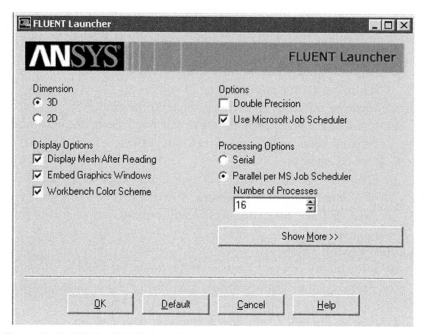

Figure 7.11 ANSYS FLUENT launcher window (https://hpc.nrel.gov/users/systems/winhpc/tips/ansys-fluent).

12. Launching FLUENT from the command line

Although the method described previously is recommended as a starting point for running FLUENT with the Microsoft Job Scheduler, further options are provided to meet your specific needs. FLUENT allows you to do any of the following with the Microsoft Job Scheduler. You can submit FLUENT jobs using the Microsoft Job Scheduler by using the −ccp flag in the FLUENT startup command. Open up a command prompt and CD to the directory where your case and data file is located and type: `fluent 3d -ccp headnode -tnprocs`.

13. Journal files

A journal file contains a sequence of FLUENT commands, arranged as they would be typed interactively into the program or entered through the GUI. The GUI commands are recorded as Scheme code lines in journal files. FLUENT creates a journal file by recording everything you type on the command line or enter through the GUI. You can also create journal files manually with a text editor. The purpose of a journal file is to automate a series of commands instead of entering them repeatedly on the command line. Another way is to produce a

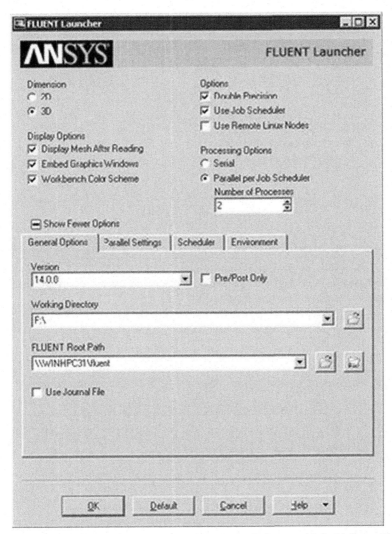

Figure 7.12 ANSYS FLUENT launcher window: expanded menu (default) (https://hpc. nrel.gov/users/systems/winhpc/tips/ansys-fluent).

record of the input to a program session for later reference, although transcript files are often more useful for this purpose. Command input is taken from the specified journal file until its end is reached, at which time control is returned to the standard input (usually the keyboard). Each line from the journal file is echoed to the standard output (usually the screen) as it is read and processed. Refer to FLUENT documentation for more information about how to create journal files.

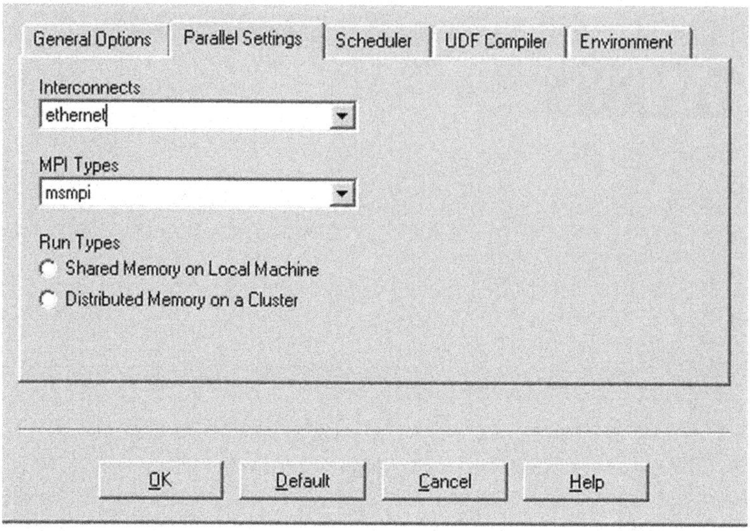

Figure 7.13 ANSYS FLUENT launcher window: parallel settings menu (https://hpc.nrel. gov/users/systems/winhpc/tips/ansys-fluent).

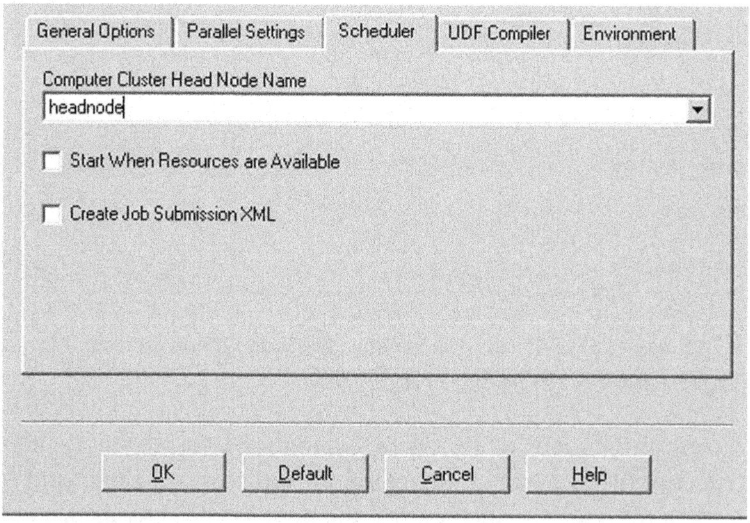

Figure 7.14 ANSYS FLUENT launcher window with Scheduler tab settings (https://hpc. nrel.gov/users/systems/winhpc/tips/ansys-fluent).

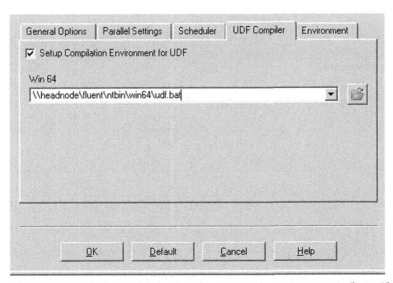

Figure 7.15 ANSYS FLUENT launcher window: UDF tab settings menu (https://hpc. nrel.gov/users/systems/winhpc/tips/ansys-fluent).

14. What are cores and sockets of nodes?
    a. **Core**: Refers to a single processing unit capable of performing computations. A core is the smallest unit of allocation available in HPC Server 2008.
    b. **Socket**: Refers to a collection of cores with a direct pipe to memory. It is the physical CPU chip. Each socket contains one or more cores. Note that this does not necessarily refer to a physical socket, but rather to the memory architecture of the machine, which will depend on your chip vendor.
    c. **Node**: Refers to an entire compute node. Each node contains one or more sockets (Figure 7.16).
    d. Specifying which nodes/sockets/cores for FLUENT runs? You can specify which nodes, sockets, or cores to run on by selecting the option on the Scheduler tab for Cores, Sockets, or Nodes (Figure 7.16).
       – **Cores** If you choose two cores, it will spawn one core on one node and one core on another.
       – **Sockets** If you choose two sockets, it will spawn two cores on one node.
       – **Nodes** If you choose two nodes, it will spawn on all cores on two available nodes. Note: No other jobs or tasks can be started on that node.

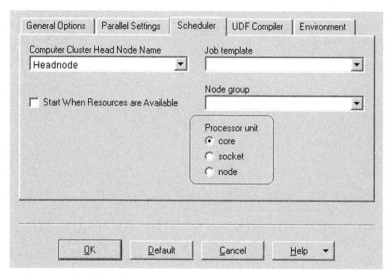

Figure 7.16 ANSYS FLUENT launcher window: cores, sockets, and nodes setting.

e. Specifying specific cores on nodes: In the Environment tab in the FLUENT GUI, add the following environment variable:

CCP NODES = %CCP NODES% -cores # (where # is the number of cores you want to use on each node) (Figure 7.17). For example, if you were to choose CCP NODES = %CCP NODES% -cores four, it would spawn four cores on each of the nodes.

f. It is sometimes confusing to know when to use cores, sockets, or nodes. In general, the rule is:

Use core allocation if the application is CPU–intensive; the more processors you can throw at it, the better. Use socket allocation if memory access is what bottlenecks your application's performance. Because the amount of data that can come from memory is what limits the speed of the job, running more tasks on the same memory bus will not result in speedup, because all of those tasks are fighting over the path to memory.

Use node allocation if some node-wide resource is what bottlenecks your application. This is the case with applications that rely heavily on access to disk or network resources. Running multiple tasks per node will not result in speedup because all of those tasks are waiting for access to the same disk or network pipe.

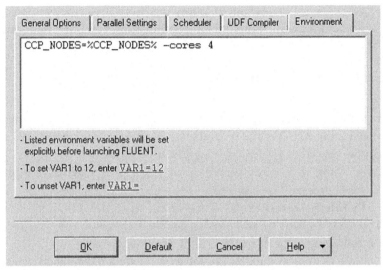

| General Options | Parallel Settings | Scheduler | UDF Compiler | Environment |

CCP_NODES=%CCP_NODES% –cores 4

· Listed environment variables will be set
  explicitly before launching FLUENT.
· To set VAR1 to 12, enter VAR1=12
· To unset VAR1, enter VAR1=

OK   Default   Cancel   Help ▾

Figure 7.17 ANSYS FLUENT launcher window: environment tab setting.

15. Some key facts
   a. The unit type set on your job also applies to all tasks in that job.
   b. You cannot have a job requesting four nodes with a bunch of tasks requesting two cores each.
   c. You can still use batch scripts or your applications mechanisms to launch multiple threads or processes on the resources your job is allocated.
   d. By using these correctly, you can improve your cluster use because jobs are more likely to get only the resources they need.

4.2 Troubleshooting FLUENT/HPC Issues

4.2.1 Windows 7: Specific Issues

1. Turning User Account Control on or off
   User Account Control can help prevent unauthorized changes to your computer. It works by prompting you for permission when a task requires administrative rights, such as installing software or changing settings that affect other users.
   a. Open User Accounts by clicking the Start button, clicking Control Panel, clicking User Accounts and Family Safety (or clicking User Accounts, if you are connected to a network domain), and then clicking User Accounts.

    b. Click Turn User Account Control on or off. If you are prompted for an administrator password or confirmation, type the password or provide confirmation.

2. Disable IPv6: IPv6 is the latest address protocol that will eventually replace IPv4. From Windows Vista onward, it has been kept enabled by default, but IPv6 is not yet common and many types of software, routers, modems, and other network equipment do not yet support it, including ANSYS. It is recommended to disable this protocol. Details can be viewed at [1] for disabling the IPv6 protocol.

3. Error: "job" is not recognized as an internal or external command, operable program, or batch file.

    Problem Description: When trying to launch FLUENT from a Windows 7 client to a Windows 2008 HPC Server R2, you receive the following error in the FLUENT window:

    "job" is not recognized as an internal or external command, operable program, or batch file.

    Explanation: Windows has a 260 character limit in the PATH variable. When installing Microsoft HPC 2008 R2 Client software, it appends the bin path to the beginning of the PATH system variable. When the 260 character limit is reached, it throws the following entry (the bin path of the HPC client software) in the FLUENT window. To verify whether this is the root cause of the error thrown, type: path in a command window. It should confirm that the output of path is missing the path to the HPC client bin directory.

    Resolution: Trim the System PATH variable so that it does not exceed the 260-character Windows limit.

4. FLUENT hangs when launching

    You launch FLUENT and the FLUENT window reports "Waiting for CCP Scheduler @headnode ..." and hangs.

    Resolution 1: The most likely reason FLUENT is hanging at this point is there is a username and/or password issue on any one of the compute nodes. The resolution is to clear the cached password and reset the password (Figure 7.18). If this is not the case, look at Resolution 2.

    Clearing the Cached Password:

    Resolution 1:

1. Open up the HPC job manager.
2. Open the Options menu.
3. Choose "Clear Cached Job Credentials."

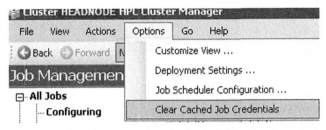

Figure 7.18 Setting cached password.

Resetting the Password

1. Open up the HPC Job Manager
2. Open up the Actions menu, Job Submission, New Job
3. Choose Task List in the left panel
4. In the command line box, type: cmd.exe
5. Choose Save, then Submit
6. You will be prompted to enter and save your password
7. Restart ANSYS FLUENT

Resolution 2: This behavior could also be caused by the order of the network bindings. Make sure that the private NIC is listed first in Adapters and Bindings.

5. An ANSYS FLUENT job seems to run indefinitely within the HPC Job Scheduler when running in batch mode

    Make sure that you have specified a working directory before launching ANSYS FLUENT and that the directory is shared. This directory must also be mapped and you must use the mapped drive letter in the working directory text box.

6. Error when trying to write out a FLUENT case or data file

    Check the order of the network bindings on the head node. The private network binding should be the first in the list (on top). The preferred network binding order is: private, Infiniband, MPI.

7. Launching FLUENT from a compute node is starting up slowly

    Sometime when ANSYS FLUENT is launched on a compute node or a client machine, it takes a long time to open. The remedy is to check the order of the NIC cards on all nodes on the cluster. The private NIC should be listed first.

8. Slow read or write times with large case and data files

    If you are experiencing slow load times when reading or writing the case and data file, check the bandwidth between the host and node 0. The command is typed in the FLUENT window: (bandwidth 0 999999 " ")

Table 7.2 Latencies and bandwidths

Interconnect	Latency (microseconds)	Bandwidth (millions of Bits per second)
GigE	50	100
Winsock direct	15	800
Transmission Control Protocol/Internet protocol	45	200
IBAL	5	1000

9. Parallel Performance Issues

For optimal FLUENT performance, low-latency is essential. Bandwidth is the term used to describe the amount of data that can be transferred over a network cable or network device in a fixed amount of time. Bandwidth is measured in bits per second or in higher units such as millions of bits per second. Latency refers to any of several kinds of delays typically incurred in processing network data. A so-called low latency network connection is one that generally experiences small delay times, whereas a high latency connection generally has long delays. Once ANSYS FLUENT is launched in the ANSYS FLUENT window, type: (bandwidth # # " "), where the first # would be the first node on cluster 0 and the second # would be the last node on the cluster you are using to run ANSYS FLUENT. For example (bandwidth 0 7 " ") can be used to measure the bandwidth and latency between nodes 0 and 7 (eight cores). It is expected that you will see similar results based on Table 7.2. If not, running the bandwidth command again using one node at a time is recommended—for example (bandwidth 0 1 " "), (bandwidth 0 2 " "), (bandwidth 0 3 " "), etc.—until the node that is experiencing a network problem is identified.

REFERENCE

[1] How to Disable Certain Internet Protocol Version 6 (IPv6) Components in Windows Vista, Windows 7 and Windows Server 2008, 2011. Website, http://support.microsoft.com/kb/929852.

CHAPTER 8

Graphics Processing Unit Technology

1. INTRODUCTION TO GRAPHICS PROCESSING UNIT

Throughout the text we have mentioned Graphics Processing Units (GPUs) but we have not gone into detail. This chapter will do so. The GPU is no longer hypothetical; Nvidia introduced it in 1999–2000. GPUs are strong central processing units (CPUs) that have the ability to perform not just graphics jobs, but also calculations. They are like CPUs but much more efficient. After the advent of GPUs, the scientific community began to use them in general-purpose computational applications, especially in the fields such as medical imaging and electromagnetics. They found the floating point performance to be excellent. However, one shortcoming was that the general-purpose graphics processing unit (GPGPU) required programming languages such as OpenGL and C. Developers had to make their applications look like graphics applications and map them into problems that draw polygons and triangles. This limited the performance accessibility of GPUs.

Nvidia realized the potential of bringing this performance to the largest scientific and research community and decided to invest in modifying the GPU to make it fully programmable for scientific applications and to provide support for higher-level languages such as C and C++. This led to the emergence of Compute Unified Device Architecture (CUDA).

To understand why GPUs are necessary and why graphics cards are useful for computational purposes, we have to go the classical Moore's law, shown in Figure 8.1. The graph indicates that the transistor count doubles every year. By the end of 2011, transistors exceeded processor chips by 2 million. Imagine how much heat would be generated if so many transistors worked together. Research and manufacturing are ongoing, extending the line to 5 million by 2020 following Moore's law, but this is impractical from an efficiency point of view. From the viewpoint of High-Performance Computing (HPC), we should not compromise on efficiency. Thus, if an increase in the transistor count causes efficiency to drop, an alternate solution is mandatory. This purpose has been fulfilled by GPUs.

Using HPC for Computational Fluid Dynamics
ISBN 978-0-12-801567-4
183

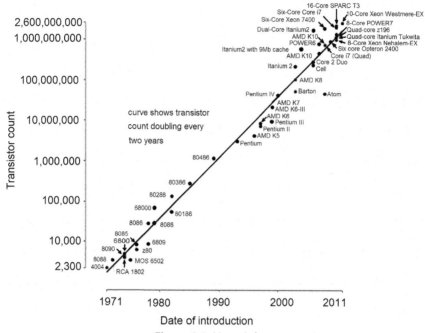

Figure 8.1 Moore's law.

Besides heating there are basic problems, such as that the power required to run does not scale as much as the processor frequency, power $\sim$ (frequency)^3. In addition, more processing power means more memory is required, which does not scale up as much as central processing unit (CPU) power. The Moore's law limitation is based on a single processor whose limit has been reached, as indicated by the clutter of machines at the top. Now, consider a single core with a 1-GHz processor and a single core with a 4-GHz processor. Then, for serial processing power $\sim$ (frequency)^3 and thus 64 W, assuming 1 W for a single core of 1 GHz.

Consider parallel cores, each of which contains a single core with a 1-GHz processor as before, but also with four core processor chips, each core of which has 1 GHz. In this way, power $\sim$ (number of cores), and thus 4 W.

GPUs are sophisticated and have been successful in gathering a market that is hungry for better and faster rendering. Their development is pushed by the supercomputing industry, which needs a high-quality product; it is inexpensive but has proven to be a real alternative for scientific applications.

2. INSIDE A GPU

The model for GPU contains a CPU and a GPU in a heterogeneous environment. The sequential part of the task is solved by the CPU while the compute-intensive part is solved by the GPU. The parallel programming is built into the GPU through parallel kernel applications. A kernel is a program that provides a bridge between hardware and software. This mapping of function to the GPUs involves rewriting the function to expose it for parallelism and to transfer the data to and from the GPU. Unlike the conventional cores in Intel or AMD processors, Nvidia GPUs are different: they are based on CUDA. CUDA contains hundreds of processors cores operating together. It is supported by the CUDA parallel programming model, which can be in any high level language such as C, C++, or FORTRAN. CUDA parallel programming guides programmers to partition the problem into coarse sub-problems that can be solved independently in parallel.

The GPU core is the stream processor and several stream processors are grouped into stream multiprocessors. They all are single instruction multiple data (SIMD), as described in Chapter 3 and given in Figure 8.2. SIMD architecture is obviously particularly efficient when the same set of operations can be applied to all data items. Interestingly, the explicit high-resolution schemes are good candidates for SIMD implementation

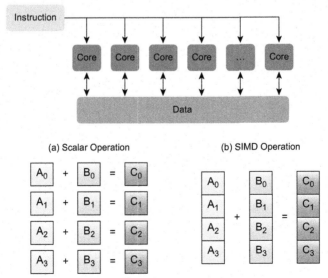

Figure 8.2 Stream multiprocessing in GPUs with (a) scalar and (b) SIMD operation.

because the update of each grid-cell involves a fixed set of given arithmetic operations and seldom introduces conditional branches into the instruction stream.

GPUs consist of many simple cores designed for simpler tasks such as high throughput (bandwidth, or the packets of data per unit time) and lower latency. Decisions regarding how to use GPUs effectively mainly depend on how to program the code to run efficiently on GPUs, such as CUDA, which is used extensively in many GPUs. It is specially designed for GPGPU computing and offers a compute-designed interface as well as explicit GPU memory managing.

A GPU chip has two parts: one acts as a host that functions as a CPU and contains multi-cores; the other is the CUDA, which act as a device. The CPU sends the data to the CUDA part for further processing. GPU programming usually runs on C++ and the programming model operates on a CPU (host) and device (GPU). In CUDA, a kernel (a program that bridges hardware and software) is executed by many threads. A thread is a sequence of executions, so in a multi-thread, many threads run at the same time. Threads are grouped into thread blocks. The programming idea is that all threads within a thread block run in the same stream multiprocessor and threads of the same block can communicate. These threads in a block form constitute a grid, as shown in Figure 8.3.

The programming model basically deals with the threads and their hierarchy (how they are grouped in thread blocks). Threads in a thread block can communicate with each other.

All threads of the same thread block are executed in the same Stream Multiprocessor (SM) at the same time. They have shared memory, so threads within a block can communicate. All of the threads of a thread block must be executed before there will be space to schedule another thread block. Hardware schedules thread blocks onto available SMs and the

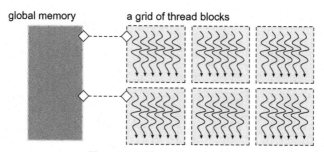

Figure 8.3 Threads and grid.

order of execution is not guaranteed. If an SM has more resources, the hardware will schedule more blocks.

Inside the SM, threads are launched in groups of 32, called warps. Warps share the control part (warp scheduler). At any time, only one warp is executed per SM. Threads in a warp will execute the same instruction. The Fermi GPU contains a maximum number of active threads of $1024 \times 8 \times 32 = 262,144$, where 1024 is the maximum threads per block, 8 is the maximum blocks per SM, and 32 is the number of SMs. Hardware separates threads of a block into warps. All threads in a warp correspond to the same thread block. The threads are placed in a warp sequentially. Figure 8.4 shows a schematic of how on the hardware the CUDA program is executed; the programming structure of CUDA is explained in the flowchart in Figure 8.5.

The structure of a GPU is similar to the structure of human cell, which has a nucleus (blocks) and the nucleus has 23 pairs of chromosomes (which we call threads here), each of which contains deoxyribonucleic acid. It is

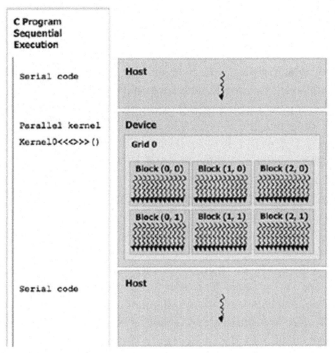

Figure 8.4 C Program execution on a GPU; wavy lines indicate threads.

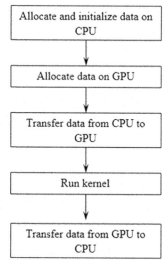

Figure 8.5 Flowchart for GPU task execution.

not clear, however, whether this logic came from the medical field or whether it is a coincidence.

2.1 GPUs Working on Three-Dimensional Meshes

To understand how GPU threads map to computational cells, consider Figure 8.6, which shows a mesh of internal cells with dimensions of nx [level] × ny[level] × nz[level]. The arrays nx[level], ny[level], and nz[level] contain the number of mesh cells in each direction and the level indicates the level in the multi-grid V-cycle. The indices of the internal cells range from $(i, j, k) = (2, 2, 2)$ to $(i, j, k) = (nx[level] + 1, ny[level] + 1, nz[level] + 1)$. The boundary cells (not shown in Figure 8.6) lie along the planes $i = 1, j = 1$, and $k = 1$ and planes $i = nx[level] + 2$, $j = ny[level] + 2$, and $k = nz[level] + 2$. The GPU grid dimensions are gx, gy, and gz, and each block has dimensions (bx, by, and bz). The GPU grid dimensions gx and gy were calculated by dividing the dimensions of the computational mesh on the current mesh level by the block size. Thus, while performing a multi-grid, GPU grid dimensions are changed to accommodate the size of the current computational mesh level. This idea is shown in the example code in Figure 8.7, in which the execution configuration in the main program (on the CPU) is changed as a function of the mesh level when calling a kernel for the GPU. This example is for the down-leg of a V-cycle, where the grid levels start at the finest level $(n = 1)$ and descend to the coarsest level $(n = ngrid)$.

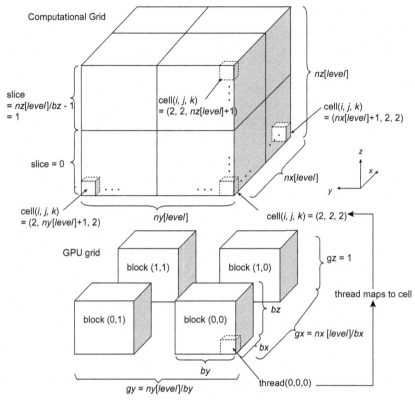

Figure 8.6 Correspondence between GPU grid and computational mesh.

Performance is sensitive to block size, so this is another area of code optimization. Block sizes must evenly divide into mesh dimensions for each mesh level for a multi-grid. A block size that can accommodate the coarsest level is selected that can accommodate all finer mesh levels. However, this may not yield optimal GPU performance because the block size is small (smaller than the warp size). A compromise between accommodating each mesh level and performance is found by using two block sizes: one for the finer meshes and one for the coarser meshes. Most computation occurs on the finer meshes (first one or two mesh levels in the V-cycle), and thus the block sizes for these levels are tuned for optimal performance.

For most problems, a good block size is (bx, by, bz) = (32, 1, 8). This can change from problem to problem, so it is best to experiment to determine the optimal sizes. For coarser meshes, a smaller block size is

```
Int main (void)
{
...
for(n = 1; n<=ngrid; n++)
{
dim3 block(bx,by,bz);
dim3 grid(nx[n]/bx,ny[n]/by);
kernel<<<grid, block>>>(...);
}
...
} // end main
__global__ void kernel(...)
{
// global thread indices
tx = threadIdx.x + blockIdx.x * blockDim.x;
ty = threadIdx.y + blockIdx.y * blockDim.y;
tz = threadIdx.z;
// convert thread indices to mesh indices
i = tx + 2;
j = ty + 2;
for(slice=0; slice<=nz[n]/blockDim.z-1;
slice++)
{
k = tz + slice * blockDim.z + 2;
m = i + (j-1)*(nx[n]+2) +
(k-1)*(nx[n]+2)*(ny[n]+2) + begin[n] - 1;
...
computations
...
} // end slice
} // end kernel
```

Figure 8.7 CUDA code.

used so that mesh resolution is evenly divisible by block size. The smaller block size delivers poor performance, but this occurs only on the coarse levels, which do not have appreciable computing times, so the effect is small.

3. NVIDIA SERIES

Nvidia is the only company in the world that is manufacturing GPUs: there is no competitor. Nvidia has the following famous products regarding GPU cards.

3.1 Nvidia G80

G80 or GeForce 8800 is a product with modified technology of the GPU computing model of Nvidia. It has several interesting features.
1. G80 was the first GPU to support C; hence, it allows programmers to use GPU without learning a new language.
2. G80 was the first GPU to use the scalar thread processor.
3. G80 introduced the SIMT algorithm, in which multiple threads are executed simultaneously using a single set of data instructions.
4. G80 introduced shared memory for inter-thread communication.

3.2 Nvidia GT200

In June 2008, Nvidia introduced a major revision to the G80 architecture. The second generation had an increased number of streaming processor cores, from 120 to 240. It had interesting features:
1. The processor register was double in size from G80, allowing a greater number of threads to execute on-chip at any given time.
2. Hardware memory coalescing was added to improve memory access.
3. Double precision floating point support was added.

3.3 Nvidia Fermi

The Fermi architecture is the most significant step forward in GPU technology. The G80 and GT200 designs were initial versions of unified graphics and parallel processing and extension in G80 technology, respectively. Nvidia employed full power in developing the architecture of Fermi by not repeating the flaws (if any) of G80 and GT200. Key areas of improvement were:
1. There was improved double precision performance.
2. Some parallel algorithms were unable to use the GPU's shared memory and users requested a true cache architecture to assist them.
3. Many CUDA programs requested more than 16 KB of shared memory.
4. Previously there had been limitations in faster context switching between application programs and graphics.

These factors compelled Nvidia to develop a GPU that not only increased computational power but improved the programmability and compute efficiency. Fermi architecture has the following features:
1. It is a third-generation streaming multiprocessor with 32 CUDA cores per processor. This is four times faster than GT200.

2. It is eight times the peak double precision floating point performance over GT200.
3. Fermi has improved memory subsystems.
4. It has 10 times faster data switching.
5. It has full 64-bit precision support.

The first Fermi GPU had 3.0 billion transistors featuring up to 512 CUDA cores. These cores are arranged in 16 SM of 32 cores each. Fermi has 64-bit memory partitions with a 384-bit memory interface; thus, it can support up to 6 GB of memory.

4. COMPUTE UNIFIED DEVICE ARCHITECTURE

The parallel programming guides programmers to partition the problem into coarse sub-problems that can be solved independently in parallel. Fine-grained parallelism can then be expressed such that each sub-problem can be solved in parallel.

4.1 CUDA Computing with GPUs

Nvidia CUDA technology influences the massively parallel processing power of Nvidia GPUs. The CUDA architecture delivers GPU computing technology to Nvidia. With the CUDA architecture, developers are achieving dramatic speedups in fields such as image processing, medical resonance imaging, graphics visualization in computer-aided design, and computer-aided engineering such as computational flow dynamics (CFD) and finite element analysis applications and real-time high-definition playback and encoding.

4.2 What Can CUDA Do?

The CUDA package includes three important applications:
1. The CUDA Driver API
2. The CUDA toolkit
3. Software Development Kit (SDK) with code examples

The CUDA toolkit is in principle a C development environment and includes the actual compiler, optimized fast Fourier transform, and Basic Linear Algebra Scale. It also includes several libraries of documentation and shared libraries for the run-time environment of CUDA programs. The SDK includes examples with source codes for matrix calculations, image convolution, pseudo number generation, and a lot more.

4.3 The CUDA Architecture

The CUDA architecture contains the following basic ingredients:

1. Parallel compute engines inside Nvidia GPUs.
2. Operating system kernel-level support for hardware initialization, configuration, etc.
3. User mode driver that provides a device-level API.
4. Instruction Set Architecture (named PTX) for parallel computing and kernels and functions.

In the CUDA programming model, compute-intensive tasks of an application are grouped into an instruction set and passed on to the GPU such that each thread core works on different data but executes the same instruction. The memory arrangement of CUDA is similar to the memory hierarchy of a conventional multiprocessor. The memory levels can be divided into three groups:

1. Local registers that allow fast arithmetic logic unit operations (level 1 [L1] cache) near the core. The cache in normal CPUs is a small memory area that stores data for faster access at a later time. It then sends data to the main memory or checks whether the data are present in the cache and then directly processes without sending it to main RAM to reduce time. The data are fetched (or searched) from L1 to L2, then L3, and then the RAM.
2. Shared memory, seen by all cores of a single multiprocessor, can be compared with a second-level cache (L2) because it provides memory closer to the processors that will be used to store data that tend to be used over time by any core. The difference in CUDA is that the programmer is responsible for managing the shared memory or GPU cache.
3. The last level in this hierarchy is global memory. It can be accessed by any processor of the GPU, but at a higher latency cost. Threads can actually perform simultaneous scatter or simultaneous gather operations if those addresses are aligned in memory.

Coalesced (combined) memory access is crucial for superior kernel performance because it hides the latency of the global memory. The challenge for a CUDA software developer is thus not only parallelization of the code, but also optimization of memory access by making the best use of shared memory and coalesced access to the global device memory. Each multiprocessor also has a read-only constant cache and texture cache. The constant cache can be used by the threads of a multiprocessor when trying to read the same constant value at the same time. Texture cache, on the other hand, is optimized for 2D spatial locality and should be preferred over global device memory when coalesced read cannot be achieved.

5. GPUs IN CFD

Although much research is ongoing for GPUs, for CFD the technology is not yet mature. Scientists are testing applications and spending time to develop CFD codes for GPUs. A lot of work is still being carried out.

To run a CFD code on GPUs, the programmer has three basic options:
1. Accelerate the existing code to be able to run on GPUs
2. Redesign the code from scratch for GPUs (a lengthy and hectic procedure)
3. Make suitable algorithms in the existing code that have the ability to run on GPUs

5.1 History of Implementing GPUs on CFD Applications

Before the arrival of CUDA, the programmer's job was to transform codes into GPU operations. One example is the work of Scheidegger et al. in 2005 [1]; who used GPU implementation of the simplified marker and cell method. The code was supposed to solve 2D incompressible Navier–Stokes (NS) with structured grid. A central difference scheme with a hybrid donor cell approach was used in the code. There was a special memory to store velocity fields, called texture memory. It was also called a pixel buffer. Jacobi iteration was performed to solve pressure Poisson equation. Two types of GPU were used in the simulation: GeForce FX 5900 (NV35) and GeForce 6800 Ultra (NV40). The CPU used was a Pentium IV with a 2-GHz processor. The CPU was better than the GPU, but only for a small mesh size. However, for larger mesh sizes the GPU was 16 times faster. Cases that were analyzed were lid-driven cavity, rising smoke at high Reynolds number (Re) and low-Re flow over a 2D car edge.

Elsen et al. in 2008 [2] used a GPU to solve incompressible NS over simple geometries and compressible Euler equations for complex geometries such as a hypersonic vehicle at Mach 5 and NACA 0012. For simple geometries, the GPU was 40 times faster than the CPU whereas it was 20 times faster for complex ones. The CPU was an Intel Core-to-Duo with a 2.4-GHz single core and the GPU was Nvidia model 8800 GTX. The code was called the Stanford University Solver; it was 3D unsteady RANS-based. A finite difference scheme with a vertex-centered solver was employed on multi-block meshes. Time discretization was performed using fifth-order

Runge–Kutta. A multi-grid solver was also used to accelerate convergence. BrookGPU language (FORTRAN-based) was used in GPU programming.

In 2008 [3]; Brandvik and Pullan used GPU to solve 2D and 3D Euler problems. The original code was FORTRAN-based to be solved on a CPU. The application area was flow over a turbine blade, which was solved in 2D for a transonic turbine cascade and with 3D secondary flow development in a low-speed linear turbine cascade. With the 2D solver and BrookGPU language, the code was supposed to be 29 times faster than a CPU and the 3D solver was to be three times faster. Applying CUDA to the 3D solver increased performance dramatically, to 16 times that of a CPU. The CPU was a 2.33-GHz Intel C2Duo with a single core. The GPU model was Nvidia 8800 GTX. The numerical scheme used was compressible Euler equations for flow modeling using the finite volume method on structured grids. Variables were stored at cell vertices. Spatial discretization was a second-order central difference scheme and temporal was first order. No multi-grid method was employed in their approach.

In other work by Brandvik and Pullan [4]; the researchers presented a 3D NS solver on multiple GPUs with MPI. The subroutines for MPI were formulated using Python language. A source-to-source compiler was developed to convert the routines into source code, thereby targeting device architectures of GPUs. This is a new approach that makes the code flexible enough to be pliable for future architectures, as well.

In 2010 [5]; Shinn et al. performed Direct Numerical Simulation (DNS) over GPUs. The numerical technique was a fractional step method with finite volume discretization for space to solve incompressible NS, and was implemented on GPUs with CUDA. The pressure Poisson equation was solved but with a multi-grid method to accelerate convergence. The problem consisted of simulating turbulent flow in a square duct (a famous case for DNS applications) at a bulk Re of 5480 with a Taylor Re (Re_τ) of 360 with a mesh size of 26.2 million cells. The purpose was twofold: to test overall GPU performance with a CPU and to test the innate capability of GPU to perform DNS, because this was the largest problem to be tested on a single chip of Tesla C1060. The salient features of this flow were captured and were in good agreement with previous data. The GPU-based solver was over an order of magnitude faster than the CPU-based version. In the same year, Chaudhary et al. [6] extended the research to include magnetohydrodynamics and studied the magnetic field effects on turbulent flow in a square duct. Direct numerical simulations were performed at a bulk Re of 5500 at different Hartmann numbers (dimensionless number used in magnetic flow studies; the ratio of electromagnetic force to viscous force) to vary the magnetic field.

For 3D implementation of an incompressible NS solver on multiple GPUs, the credit goes to Thibault and Senocak [7]; who worked on it in 2009. Using CUDA, they employed a fractional-step procedure to solve the equations and the pressure Poisson equation was solved using Jacobi iteration without a multi-grid scheme. The spatial terms were discretized with second-order accurate central differences and an explicit, first-order accurate Euler scheme was used for temporal advancement. Validation of GPU implementation and assessment of speedup were performed by solving the problem of laminar flow in a lid-driven cavity.

A simple problem of lid-driven cavity flow was analyzed. The Re was 1000. The famous lid-driven cavity problem that was initially tested by Ghia. Different mesh sizes were tested and analyzed based on the performance over a GPU and serial CPU. The serial CPUs were an Intel Core 2 Duo 3.0 GHz and an AMD Opteron 2.4 GHz (quad core). Kernels were used for the velocity predictor step and solution of the Poisson equation was implemented using the shared memory of the device. Separate kernels were dedicated for calculating the divergence field and velocity corrections using global memory. This combination of the two sets of kernels resulted in a speedup of two.

Overall performance for the solution of the incompressible problem with a Nvidia S870 T server was 100 times faster than a CPU. The speedup number was measured relative to the serial CPU version of the CFD code that was executed using a single core of an AMD Opteron 2.4-GHz processor. However, with respect to a single core of an Intel Core 2 Duo 3.0-GHz processor, speedup of 13 and 21 with single and dual GPUs (Nvidia Tesla C870), respectively, was obtained. The numerical methods used in CPU and GPUs were identical. It was found that multi-GPU scaling and speedup results improved with increasing computational problem size, indicating that computationally big problems can easily be tackled with GPU clusters with multi-GPUs in each node. One CPU core should be dedicated to each active GPU to obtain good scaling performance across multi-GPUs.

There has been development in multiphase application on GPUs. Griebel and Zaspel [8] implemented a two-phase NS solver on a GPU. They used a level set technique and a fractional step method to solve NS equations. The solver was implemented on multiple GPUs and data communication between GPU and CPU was accomplished through MPI. To minimize the overhead from data communication, they used the asynchronous communication feature of CUDA, in which data can be copied while computations are performed. This effectively reduced the communication time.

Everett et al. [9] carried out research investigating the use of a cluster of GPUs for large-scale CFD problems and showed an order of magnitude increase in performance and performance-to-price ratio. The research was twofold: First, they developed a 2D Euler code compressible flow solver and verified the results with reference multi-block code MBFLO. Later, modifications were performed in FORTRAN-base code of MBFLO by adding subroutines for GPU. With an eight-node cluster and 16 Nvidia 9800 GX2 GPUs, the speedup was 496 times faster than the serial CPU. An interesting aspect was the discussion of the bottlenecks in conventional solvers. Conventional solvers are usually RANS-based methods and give 1–3% accuracy. This requires a lot of computational time to complete, with a dramatic increase in computational cost. The second bottleneck is large-scale parallelization, an obvious requirement for large problems. Therefore, with the advent of GPUs, a small investment is needed that smoothes the problems of time-averaging, unsteady simulations. A small additional cost may then be added for detached eddy simulations or large eddy simulations. The speedups for 16 GPUs were also calculated and performance increased 88 times over a single CPU.

Cohen worked on developing a second-order double-precision finite volume Boussinesq code to run with CUDA architecture [10]. The code was validated on a number of Rayleigh-Bénard convection problem. Rayleigh-Bénard convection is instability of a fluid layer confined between two parallel plates; the lower plate is heated to produce a fixed temperature difference. Because liquids typically have positive thermal expansion co-efficient, the hot liquid at the bottom of the cell expands and produces an unstable density gradient in the fluid layer. If the density gradient is sufficiently strong, the hot fluid will rise, causing a convective flow that results in enhanced transport of heat between the two plates. The results were compared by developing FORTRAN-based code running on eight CPU cores. It was then compared with CUDA core performance and performance was eight times faster than two quad-core CPUs. The simulation can be run on a grid size of 3842×192 at 1.6 s/time step on a machine with a single GPU. It was concluded that the speedup gained with GPUs was the result of high bandwidth from the GPU processors to GPU memory. The GPU-to-device memory bandwidth is approximately an order of magnitude greater than CPU-to-host memory bandwidth. Among the key findings was that GPU-to-device memory bandwidth was an order of magnitude higher than CPU-to-host memory. There was one bottleneck of bandwidth latency that occurred because large datasets do not get stored even in L3 cache. According to the author of the report, the data are less calculation intensive and thus dependent on memory bandwidth. Another finding was related to the difference between single and double precision. In single precision, floating points operations do not become saturated, so double precision demands more memory. Thus, in case of bandwidth-limited applications (to smooth less compute-intensive solvers) GT 200 is supposed to perform better with a double-precision problem. The report concluded that for optimal performance of a GPU memory system, serial bottlenecks must be removed, which has not yet been done.

6. COMPUTATIONAL FLOW DYNAMICS CODING ON GPUs

The main objective of the GPU is to render geometrical primitives (points, lines, triangles, and quads), possibly combined with one or more image data as discrete pixels in a frame buffer (screen or off-screen buffer). When a primitive is rendered, the rasterizer (the form understandable to video devices such as video monitors, liquid crystal displays, printers, etc.) samples the primitive at points corresponding to the pixels in the frame buffer. Attributes such as texture coordinates are set by the application for each vertex and the rasterizer calculates an interpolated value of these attributes for each point. These bundles of values thus contain all of the information necessary to calculate the color of each point, and are called fragments (Figure 8.8). We see that whereas a **CPU** is **instruction-driven**, the programming model of a **GPU** is **data-driven**. This means that individual data elements of the data stream can be pre-fetched (taken out) from memory before they are processed, thus avoiding the memory latency that tends to block CPUs.

Graphics pipelines contain two programmable parts: the vertex processor and the fragment processor. The vertex processor operates on each vertex without knowledge of surrounding vertices or the primitive to which it belongs. The fragment processor works in the same manner: it operates on each fragment in isolation. Both processors use several parallel pipelines and thus work on several vertices/fragments simultaneously. The programs need to be implemented using both vertex and fragment processing. One or more shaders are introduced at this stage; these are codes written at a high level, often C-like, Cg, or Open GL Shading Language, and are read and compiled by the graphical application into programs run

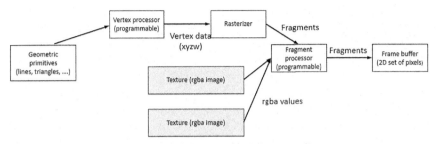

Figure 8.8 Schematic of graphics pipeline.

by the vertex and fragment processors. A description of these languages was provided in detail by Refs [11] and [12]. The compiled shaders are executed for each vertex/fragment in parallel.

6.1 Some Problems Substantially Solved on GPUs

Trond et al. [13] discussed the Lax Friedrich (LF) and Lax scheme in detail. This book does not discuss these schemes in detail, but a few highlights are provided to understand implementation on GPUs. Interested readers may refer to the famous texts on these schemes, such as Tannehill et al. [14], Anderson [15]; and Hoffmann and Chiang [16]. The schemes are conservative and can be written in one spatial dimension:

$$Q_i^{n+1} = Q_i^n - r\left(F_{i+1/2} - F_{1-i/2}\right) \tag{8.1}$$

where $Q = [\rho, \rho u, \rho v, E]^t$ and $F = F(Q) = [\rho u, \rho u^2 + p, u(E+p)]^t$ and E is the energy term indicated as $E = 1/2\,\rho u^2 + p/\gamma - 1$ and here $Q_i^n = Q_i(n\Delta t)$ and $r = \Delta t/\Delta x$. The Lax Wendroff (LW) statement says that if a sequence of approximations computed by a consecutive and consistent numerical method converges to some limit, the limit is a weak solution, assuming that $Q(x, n\Delta t)$ is a piecewise constant and $Q_i^n = Q_i(n\Delta t)$ inside each grid cell. Then, since in hyperbolic system of equations, all perturbations run with finite speed, the problem, is estimating a solution to the equation at the cell interface, which is a mathematical form of discontinuity, can be modeled as set of locally defined Initial Value Problem, IVP, on the above equation:

$$Q_t + F(Q)_x = 0,$$

$$Q(x, 0) = \begin{cases} Q_L & \text{if } x < 0 \\ Q_R & \text{if } x > 0 \end{cases}$$

This is referred to as the Riemann problem.

6.1.1 Lax Friedrich Scheme

The LF scheme takes the average of flux values at right and left boundaries. This is given with central approximation as

$$F\left[Q_i\left(x_{i-1/2}, n\Delta t\right)\right] = 1/2\left[F\left(Q_i^n\right) + F\left(Q_{i+1}^n\right)\right] \tag{8.2}$$

Then, the convective term at the next time level can be evaluated as:

$$Q_i^{n+1} = Q_i^n - \frac{1}{2}r\left[F\left(Q_{i+1}^n\right) - F\left(Q_{i-1}^n\right)\right] \tag{8.3}$$

Then, the artificial viscosity is added to make the scheme stable. This also occurs with the CFL condition, whose mathematical details are not mentioned here. The condition simply states that the domain of dependence for the exact solution, which is bounded by the characteristics of the smallest and largest eigenvalues, should be contained in the domain of dependence for the discrete equation.

The LF scheme is robust. This is largely because of the added dissipation, which tends to smear out discontinuities. A formal Taylor analysis of a single time step shows that the truncation error of the scheme is of second order in the discretization parameters. Rather than truncation errors, we are interested in the error at a fixed time (using an increasing number of time steps) and must divide by Δt, giving an error of first order. For smooth solutions, error measured in, e.g., the L1 or L1-norm therefore decreases linearly as the discretization parameters are refined, and hence we call the scheme first-order accurate.

6.1.2 Lax Wendroff Scheme

To achieve formal second-order accuracy, we modify the forward Euler method in a domain $\mathbb{R} = \left[x_{i-\frac{1}{2}}, x_{i+\frac{1}{2}}\right]$, which is given as:

$$\frac{d}{dt} \int_{x-\frac{1}{2}}^{x+\frac{1}{2}} Q(x, t)dt = F\left(Q\left(x_{i+\frac{1}{2}}, t\right)\right) - F\left(Q\left(x_{i-\frac{1}{2}}, t\right)\right) \tag{8.4}$$

Now we can obtain a set of ordinary differential equations for the cell averages $Q_i(t)$:

$$\frac{d}{dt}Q_i(t) = -\frac{1}{\Delta x_i}\left[F_{i+\frac{1}{2}}(t) - F_{i-\frac{1}{2}}(t)\right] \tag{8.5}$$

where the fluxes across the cell boundaries are given by:

$$F_{i\pm\frac{1}{2}}(t) = F\left(Q\left(x_{i\pm\frac{1}{2}}, t\right)\right) \tag{8.6}$$

and is used for integration by the midpoint method; that is,

$$Q_i^{n+1} = Q_i^n - r\left(F_{i+1/2}^{n-1/2} - F_{i-1/2}^{n+1/2}\right) \tag{8.7}$$

Then, the flux is $F_{i\pm1/2}^{n\pm1/2}$ is at the midpoint of time intervals; that is, $t^{n\pm1/2} = (n \pm 1/2)\Delta t$. Acceptable accuracy can be obtained with grid spacing $\Delta x/2$. The above equation is modified as:

$$Q_{i\pm1/2}^{n\pm1/2} = \frac{1}{2}\left(Q_{i+1}^n + Q_i^n\right) - \frac{1}{2}r\left(F_{i+1}^n - F_i^n\right) \tag{8.8}$$

Trond et al. [13] showed an example by considering the solution of a linear advection equation and Burgers equation and compared the results of LF and LW schemes. The solution with the LF scheme at time $t = 1$ s smoothed the non-smooth parts of the solution, whereas the LW scheme gave spurious oscillations, as is obvious for second-order schemes. Overall approximations were well in agreement with the analytical solution.

A composite scheme was also discussed, consisting of three steps of LW followed by the LF smoothing step. The idea is to dampen the spurious oscillations created by the LW scheme through the LF scheme, which has the numerical dissipation in it.

6.1.3 Comparison of Performance of GPU and CPU

Trond compared GPU and CPU performance [13]. The CPU was a Dell Precision G70 with EM64T dual Intel Xeon 2.8 GHz and 2 GB RAM. The operating system was Fedora. The code for the LF scheme was written in C language with the compiler icc−03−ipo−xp, version 8.1. Two GPUs were used: one was a GeForce 6800 Ultra, released in April 2004. It had Nvidia forceware version 71.8 beta drivers. It had 16 parallel pipe linings, each capable of processing vectors of length four simultaneously. The length of a vector matrix is the number of elements in the vector matrix. The other was a GeForce 7800 GTX card released in June 2005. It had a Nvidia forceware 77.72 driver and 24 pipe linings. The grids for solving the CFD problem consisted of 1024 × 1024 cells. Trond also mentioned some examples, which are explained here with reference to performance.

1. The first was a dam break problem. The assumption was the absorbing boundary condition. The solution of an expanding shockwave was found. Within the shock the rarefaction wave travels that transports water from the original deep region out to the shock. The numerical

domain consisted of a $[-1,1] \times [-1,1]$ domain with absorbing boundary conditions. These boundary conditions allow the waves to pass out of the domain without creating reflections. The simplest approach is to extrapolate grid values from inside the domain in each spatial direction.

A first-order LF scheme was used that smeared both the leading shock and the tail of the inwardly collapsing rarefaction wave. The scheme used a grid of 128×128 cells.

The thing to notice here is that the 7800 GTx is approximately two to three times faster than the 6800 Ultra. The reason is that for LF schemes the number of arithmetic operations or fragment or rendering passes is approximately equal to the arithmetic operations per grid points, which are less than the texture fetches. The number of texturing fetches per rendering pass is lower than the cost of switching between rendering buffers. Surely, the cost of switching is a major reason for this performance, which was substantially reduced in GTX owing to improved hardware drivers.

2. The second problem discussed by Troud was of bubble–shock interaction [13]. The problem considered the interaction of planar shock in air with circular region of low density. The outer radius was 0.2 with density and pressure = 1. The incoming shockwave started at $x = 0$ and propagated in a positive x-direction. The pressure behind the shock was 1D, giving a 2.95 Mach shock. The LF scheme was applied and captured the leading shock on all of the grids, whereas weaker waves were solved on the two finest grids.

For the GPU computation, the domain reduced to $[0,1] \times [0,1]$. The final simulation time was 0.2 s. The run time per time step and speedup factor are reported in Table 8.2.

Compared with the circular dam break case in Table 8.1, the run time per time step on the CPU increased by a factor between 1.3 and 1.4; the run time on the GPU increased at most by a factor of 1.1. The resulting increase in speedup factor is thus slightly less than 4/3. This can be explained as follows: Because the shallow-water equations have only three components (x, y, and t), the GPU exploits only three quarters of its processing capabilities in each vector operation, whereas the Euler equation has four components and can exploit the full vector processing capability. However, because the flux evaluation for the Euler equations is slightly more costly in terms of vector operations, and because not all operations in the shaders are vector operations of length four, the increase in speedup factor is expected to be slightly less than 4/3.

Table 8.1 Runtime per time step in seconds and speedup factor for the CPU versus the GPU implementation of LF scheme for the circular dam break problem run on a grid with N × N grid cells, for the Nvidia GeForce 6800 Ultra and GeForce 7800 GTX graphics cards

N	CPU	GF 6800	Speedup	GF 7800	Speedup
128	2.22e-03	7.68e-04	2.9	2.33e-04	9.5
256	9.09e-03	1.24e-03	7.3	4.59e-04	19.8
512	3.71e-02	3.82e-03	9.7	1.47e-03	25.2
1024	1.48e-01	1.55e-02	9.5	5.54e-03	26.7

6.2 Solving an Incompressible Flow 3D Problem

Vanka et al. [17] solved the momentum and energy equations using an explicit algorithm with second-order temporal and spatial accuracy. The finite volume method with a staggered grid was used. The momentum equations were updated explicitly; no iterations were required, and thus there were no recursive steps. The most time-consuming step was the pressure Poisson equation, which was fully implicit. The pressure Poisson equation requires convergence to a high degree and consumes nearly 80% of the total time. Successive over-relaxation (SOR) was used to solve the implicit equation, which has a better convergence rate than a pure explicit Jacobi scheme. Acceleration of convergence on a fine grid was accomplished using geometric multi-grid on a structured grid. Several levels of finite volume grids nested within a fine grid were used. Because the traditional SOR is not parallelizable, we have used a red–black coloring scheme [18] to separate the unknowns into two independent subsets. The mesh is colored like a checkerboard and first the red cells are updated, then the black cells (or vice versa). The multi-grid is implemented with a V-cycle and consists of restriction, relaxation, and prolongation. The solution of the pressure Poisson equation is done to a high accuracy, typically three or four orders of magnitude reduction in error at every time step.

Table 8.2 Run time per time step in seconds and speedup factor for the CPU versus GPUs GF 6800 and GF 7800

No of cores	CPU time	GF 6800	Speedup	GF 7800	Speedup
128	3.11E-03	7.46E-04	4.2	2.50E-04	12.4
256	1.23E-02	1.33E-03	9.3	5.56E-04	22.1
512	4.93E-02	4.19E-03	11.8	1.81E-03	27.2
1024	2.02E-01	1.69E-02	12	6.77E-03	29.8

Table 8.3 Performance for simulation of laminar flow in a lid-driven cavity

Mesh	CPU time (seconds)	GPU time (seconds)	Speedup (CPU/GPU)
$16 \times 16 \times 16$	0.46	0.34	1.35
$32 \times 32 \times 32$	4.49	0.82	5.48
$64 \times 64 \times 64$	46.15	2.84	16.25
$128 \times 128 \times 128$	420.20	17.38	24.18

Times were taken for the first 100 time steps of simulation [17].

The decrease in memory access time is interesting, as well. In the SOR scheme, textures were used to fetch the pressure data from global memory, which decreased the overall execution time by 10%.

6.2.1 Performance Measurement

The performance of the solver on a CPU (written in FORTRAN) versus on a GPU (written in CUDA) is compared for two different problems in Tables 8.1 and 8.2. The CPU was a 2.6-GHz AMD Phenom quad-core processor (single core used) and the GPU was a Tesla C2070 (Fermi architecture). The CUDA 3.2 compiler was used for the GPU executables and the GFORTRAN compiler with optimization for the CPU executables. Table 8.3 shows the simulation performance of laminar flow in a lid-driven cavity at an Re of 1000 based on the lid speed and cavity edge length. Significant performance increase has been observed with GPU. Table 8.4 shows the simulation performance of DNS of turbulent flow in a square duct at an Re of 360 based on the friction velocity and hydraulic diameter. It can be seen that large benefit is obtained with higher mesh size.

7. FINAL REMARKS

From one point of view, the GPU can be considered a parallel computer of the SIMD type, except that for the GPU there is no expressed communication

Table 8.4 Performance for DNS of turbulent flow in a square duct

Mesh	CPU time (seconds)	GPU time (seconds)	Speedup (CPU/GPU)
$128 \times 32 \times 32$	27.63	2.03	13.61
$256 \times 64 \times 64$	275.96	12.76	21.63
$512 \times 64 \times 64$	569.04	24.53	23.20
$512 \times 128 \times 128$	1997.05	97.26	20.53

Times were taken for the first 100 time steps of simulation [17].

between the nodes and pipelines anywhere in the code. From a user's point of view, the main attraction of GPUs is the price-to-performance ratio (or the ratio of price versus power consumption).

Nvidia realized the potential of bringing this performance to the largest scientific and research community and decided to invest in modifying the GPU to make it fully programmable for scientific applications and to provide support for higher-level languages such as C and C++. Thus, it can be said that GPUs are the future processors. Moreover, GPUs have successfully been implemented in CFD with advanced Nvidia generations and high-level C programming.

REFERENCES

[1] Scheidegger CE, Comba JLD, da Cunha RD. Practical CFD simulations on programmable graphics hardware using SMAC. Comput Graph Forum 2005;24(4): 715–28.

[2] Elsen E, LeGresley P, Darve E. Large calculation of the flow over a hypersonic vehicle using a GPU. J Comput Phys 2008;227(24):10148–61.

[3] Brandvik T, Pullan G. Acceleration of a 3D Euler solver using commodity graphics hardware. In: 46th AIAA aerospace sciences meeting; 2008.

[4] Brandvik T, Pullan G. An accelerated 3D navier-stokes solver for flows in turbo-machines. In: ASME Turbo Expo 2009; 2009.

[5] Shinn AF, Vanka SP, Hwu WW. Direct numerical simulation of turbulent flow in a square duct using a graphics processing unit (GPU). In: 40th AIAA fluid dynamics conference; 2010.

[6] Chaudhary R, Vanka SP, Thomas BG. Direct numerical simulations of magnetic field effects on turbulent flow in a square duct. Phys Fluids 2010;22(7):1–15.

[7] Thibault J, Senocak I. CUDA implementation of a Navier-Stokes solver on multi-GPU desktop platforms for incompressible flows. In: 47th AIAA Aerospace Sciences Meeting; 2009.

[8] Griebel M, Zaspel P. A multi-GPU accelerated solver for the three-dimensional two-phase incompressible Navier-Stokes equations. Comput Sci Res Dev 2010; 25(1–2):65–73.

[9] Phillips, EH. Rapid aerodynamic performance prediction on a cluster of graphics processing units. AIAA Paper, In: Proceedings of the 47th AIAA Aerospace Sciences Meeting, AIAA 2009-565, 2009.

[10] Cohen JM, Molemaker MJ. A fast double precision CFD code using CUDA. In: 21st International conference on parallel computational fluid dynamics; 2009.

[11] Fernando R, Kilgard MJ. The Cg tutorial: The definitive guide to programmable real-time graphics. Addison-Wesley longman publishing Co., Inc.; 2003.

[12] Rost RJ. OpenGLR shading language. Addison Wesley Longman Publishing Co., Inc.; 2004.

[13] Trond RH, et al. How to solve systems of conservation laws numerically using the graphics processor as a high-performance computational engine. http://www.sintef. no/globalassets/upload/ikt/9011/geometri/gpgpu/pdf/conslaws-gpu.pdf. (accessed in December 2014).

[14] Tannehill CJ, et al. Computational Fluid Mechanics and Heat Transfer. Taylor and Francis.

[15] Anderson JD. Computational fluid dynamics-the basics with applications. New York: McGraw-Hill; 1995.
[16] Hoffmann KA, Chiang ST. Computational fluid dynamics. fourth ed., vol. I. USA: EES; 2000.
[17] Vanka PS, et al. Computational fluid dynamics using graphics processing units: challenges and opportunities. In: Proceedings of the ASME 2011 international mechanical engineering congress & exposition IMECE2011, November 11–17, 2011, Denver, Colorado, USA, IMECE2011-65260.
[18] Yavneh I. On red black SOR smoothing in multigrid. SIAM J Sci Comput 1994; 17(1):180–92.

INDEX

Note: Page numbers followed by "f" and "t" indicate figures and tables respectively.

Printed in the United States
By Bookmasters